STUDY AND REVISION GUIDE

Cambridge IGCSE™

Biology

Third Edition

Dave Hayward

HODDER
EDUCATION
AN HACHETTE UK COMPANY

Every effort has been made to trace all copyright holders, but if any have been inadvertently overlooked, the Publishers will be pleased to make the necessary arrangements at the first opportunity.

Although every effort has been made to ensure that website addresses are correct at time of going to press, Hodder Education cannot be held responsible for the content of any website mentioned in this book. It is sometimes possible to find a relocated web page by typing in the address of the home page for a website in the URL window of your browser.

Hachette UK's policy is to use papers that are natural, renewable and recyclable products and made from wood grown in well-managed forests and other controlled sources. The logging and manufacturing processes are expected to conform to the environmental regulations of the country of origin.

Orders: please contact Hachette UK Distribution, Hely Hutchinson Centre, Milton Road, Didcot, Oxfordshire, OX11 7HH. Telephone: +44 (0)1235 827827. Email education@hachette.co.uk Lines are open from 9 a.m. to 5 p.m., Monday to Friday. You can also order through our website: www.hoddereducation.com

ISBN: 978 1 3983 6134 8

© Dave Hayward 2022

First published in 2005
Second edition published in 2016
This edition published in 2022 by
Hodder Education,
An Hachette UK Company
Carmelite House
50 Victoria Embankment
London EC4Y 0DZ

www.hoddereducation.co.uk

Impression number 10 9 8 7 6 5 4 3 2 1

Year   2026 2025 2024 2023 2022

Cover photo © Werner Dreblow / stock.adobe.com

Typeset in India

Printed in Spain

A catalogue record for this title is available from the British Library.

Contents

Answers to exam-style questions are available at:
www.hoddereducation.co.uk/cambridgeextras

© Dave Hayward 2022

Introduction

Welcome to the Cambridge IGCSE™ Biology Study and Revision Guide. This book has been written to help you revise everything you need to know and understand for your exam. Following the Biology syllabus, it covers all the key Core and Extended content, and provides sample questions and answers, as well as practice questions, to help you learn how to answer questions and to check your understanding.

How to use this book

Key objectives

The key skills and knowledge covered in the chapter. You can also use this as a checklist to track your progress.

Skills

Key practical skills coverage will help you to consolidate your understanding of practical work you have undertaken in your lessons, and to describe and evaluate these skills effectively.

Key mathematical skills are covered to help you to demonstrate these skills correctly.

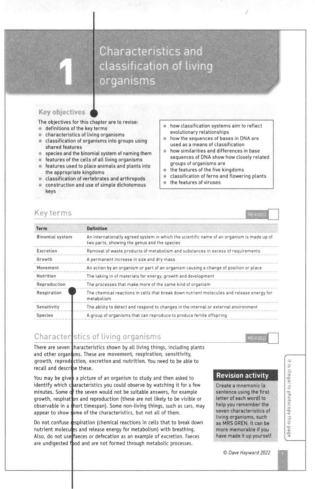

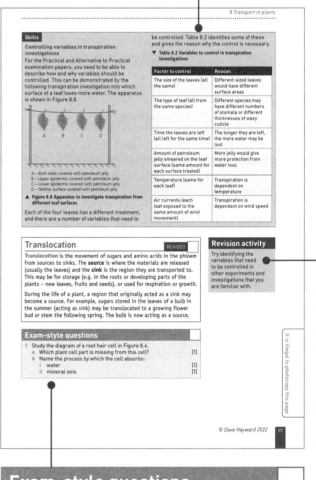

Key terms
Definitions of key terms from the syllabus that you need to know.

Exam-style questions
Practice questions, set out as you would see them in the exam paper, for you to answer so that you can check what you have learned.

Revision activities
Examples of strategies to help you revise effectively.

Sample questions

Exam-style questions for you to think about.

Student's answers

Typical student answers to see how the question might have been approached.

Correct answers

Model student answers, based on the teacher's comments on the typical student answers.

Extended syllabus

Content for the Extended syllabus (Supplement material) is shaded yellow.

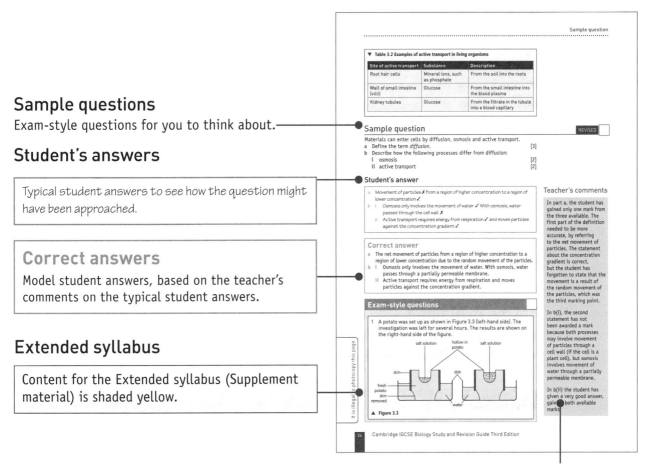

Teacher's comments

Feedback from a teacher, showing what was good, and what could have been improved.

Answers

Worked answers to the Exam-style questions can be found at **www.hoddereducation.co.uk/cambridgeextras**.

Exam breakdown

You will take three examinations at the end of your studies. If you have studied the Core syllabus content you will take Paper 1 and Paper 3, and either Paper 5 or Paper 6. If you have studied the Extended syllabus content (Core and Supplement) you will take Paper 2 and Paper 4, and either Paper 5 or Paper 6.

Paper 1: Multiple choice (Core)	Paper 3: Theory (Core)
45 minutes	1 hour 15 minutes
40 marks	80 marks
40 four-option, multiple-choice questions based on the Core subject content	Short-answer and structured questions based on the Core subject content
30% of your grade	50% of your grade

Paper 2: Multiple choice (Extended)	Paper 4: Theory (Extended)
45 minutes	1 hour 15 minutes
40 marks	80 marks
40 four-option, multiple-choice questions, based on the Core and Supplement subject content	Short-answer and structured questions, based on the Core and Supplement subject content
30% of your grade	50% of your grade

Paper 5: Practical test	Paper 6: Alternative to practical
1 hour 15 minutes	1 hour
40 marks	40 marks
Questions will be based on the experimental skills in Section 4	Questions will be based on the experimental skills in Section 4
20%	20%

Examination terms explained

The examination syllabus gives a full list of the command terms used in the exam and how you are expected to respond. This is summarised below.

Command word	Explanation
Calculate	Work out from given facts, figures or information
Compare	Identify/comment on similarities and/or differences
Define	Give the precise meaning
Describe	State the points of a topic / give the characteristics and main features
Determine	Establish an answer using the information available
Evaluate	Judge or calculate the quality, importance, amount or value of something
Explain	Set out purposes or reasons / make the relationships between things evident / state why and/or how, and support with relevant evidence
Give	Produce an answer from a given source or use recall/memory
Identify	Name/select/recognise
Outline	Set out the main points briefly
Predict	Suggest what may happen, based on available information
Sketch	Make a simple freehand drawing, showing the key features, and taking care over proportions
State	Express in clear terms
Suggest	Apply knowledge and understanding to situations where there is a range of valid responses, in order to make proposals / put forward considerations

1 Characteristics and classification of living organisms

Key objectives

The objectives for this chapter are to revise:
- definitions of the key terms
- characteristics of living organisms
- classification of organisms into groups using shared features
- species and the binomial system of naming them
- features of the cells of all living organisms
- features used to place animals and plants into the appropriate kingdoms
- classification of vertebrates and arthropods
- construction and use of simple dichotomous keys

- how classification systems aim to reflect evolutionary relationships
- how the sequences of bases in DNA are used as a means of classification
- how similarities and differences in base sequences of DNA show how closely related groups of organisms are
- the features of the five kingdoms
- classification of ferns and flowering plants
- the features of viruses

Key terms

REVISED

Term	Definition
Binomial system	An internationally agreed system in which the scientific name of an organism is made up of two parts, showing the genus and the species
Excretion	Removal of waste products of metabolism and substances in excess of requirements
Growth	A permanent increase in size and dry mass
Movement	An action by an organism or part of an organism causing a change of position or place
Nutrition	The taking in of materials for energy, growth and development
Reproduction	The processes that make more of the same kind of organism
Respiration	The chemical reactions in cells that break down nutrient molecules and release energy for metabolism
Sensitivity	The ability to detect and respond to changes in the internal or external environment
Species	A group of organisms that can reproduce to produce fertile offspring

Characteristics of living organisms

REVISED

There are seven characteristics shown by all living things, including plants and other organisms. These are **movement**, **respiration**, **sensitivity**, **growth**, **reproduction**, **excretion** and **nutrition**. You need to be able to recall and describe these.

You may be given a picture of an organism to study and then asked to identify which characteristics you could observe by watching it for a few minutes. Some of the seven would not be suitable answers, for example growth, respiration and reproduction (these are not likely to be visible or observable in a short timespan). Some non-living things, such as cars, may appear to show some of the characteristics, but not all of them.

Do not confuse respiration (chemical reactions in cells that break down nutrient molecules and release energy for metabolism) with breathing. Also, do not use faeces or defecation as an example of excretion. Faeces are undigested food and are not formed through metabolic processes.

Revision activity

Create a mnemonic (a sentence using the first letter of each word) to help you remember the seven characteristics of living organisms, such as MRS GREN. It can be more memorable if you have made it up yourself.

Sample question

Name three characteristics of living things that you would expect an organism to show, other than irritability. [3]

Student's answer

> Movement ✓, reproduction ✓ and sensitivity ✗

Teacher's comments

> The first two answers are fine. However, the term sensitivity means the same as irritability, which has already been given in the question, so it did not earn a mark. Other possible answers are respiration, growth, excretion and nutrition.

Classification systems

Classification makes the identification of living organisms easier – there are more than one million different **species** already identified. It involves sorting organisms into groups according to the features they have in common. The biggest group is called a kingdom. Each kingdom is divided into smaller groups, which include genus and species. Organisms can exist in only one group at each level of classification. For example, an organism can belong to only one kingdom or one genus.

When learning details about the classification of an organism, remember to identify what features are adaptations to its environment.

Binomial nomenclature

The **binomial system** is a worldwide system used by scientists. The scientific name of an organism is made up of two parts – genus and species – which are in Latin. The genus always has a capital letter – for example, *Panthera leo* is the binomial name for lion.

Dichotomous keys

Keys are often used by biologists in the process of identifying organisms. You need to be able to construct and use a dichotomous key, i.e. a key that branches into two at each stage, requiring you to choose between alternatives.

When completing a question involving a dichotomous key, make sure you work through the key properly to select your answer, rather than jumping to a statement that appears to fit the organism.

Skills

Construction of dichotomous keys

You need to be able to develop the skill of constructing simple dichotomous keys, based on easily identifiable features. If you know the main characteristics of a group, it is possible to draw up a systematic plan for identifying an unfamiliar organism. The first question should be based on a feature that will split the group into two.

The question is going to generate a 'yes' or 'no' answer. For each of the two subgroups formed, a further question based on the features of some of that subgroup should then be developed. This questioning can be continued until every member of the group has been separated and identified.

Classification and evolutionary relationships

By classifying organisms, it is also possible to understand evolutionary relationships. Classification is traditionally based on studies of **morphology** (the study of the form, or outward appearance, of organisms) and **anatomy** (the study of their internal structure, as revealed by dissection). Vertebrates all have a vertebral column, a skull protecting a brain and a pair of jaws (usually with teeth). By studying the anatomy of different groups of vertebrates, it is possible to gain an insight into their evolution.

Use of DNA sequencing in classification

The sequences of DNA and of amino acids in proteins are used as a more accurate means of classification than studying morphology and anatomy. Eukaryotic organisms contain chromosomes, made up of strings of genes. Genes are made of DNA, which is composed of a sequence of bases (see Chapter 4). Each species has a distinct number of chromosomes and a unique sequence of bases in its DNA, making it identifiable and distinguishable from other species.

The process of biological classification involves organisms being grouped together according to whether or not they have one or more unique characteristics in common derived from the group's last common ancestor, which are not present in more distant ancestors. Organisms that share a more recent ancestor (and are more closely related) have DNA base sequences that are more similar than those that share only a distant ancestor.

Features of organisms

REVISED

The cells of all living organisms contain cytoplasm, a cell membrane and DNA as genetic material. Two kingdoms are the plant and animal kingdoms.

Plants are made up of many cells – they are multicellular. Plant cells have an outside wall made of cellulose. Many of the cells in plant leaves and stems contain chloroplasts with photosynthetic pigments, such as chlorophyll. Plants make their food through photosynthesis.

Animals are multicellular organisms whose cells have no cell walls or chloroplasts. Most animals ingest solid food and digest it internally.

For the core syllabus, you only need to learn the main groups of vertebrates and arthropods.

Classification of vertebrates

Vertebrates are animals with backbones (part of an internal skeleton). Vertebrates are divided into five groups called classes. Details of each group are given in Table 1.1. You only need to be able to describe visible external features, but other details can be helpful (see the 'Other details' column).

▼ **Table 1.1 Classification of vertebrates**

Vertebrate class	Body covering	Movement	Reproduction	Sense organs	Other details	Examples
Fish	Scales	Fins (also used for balance)	Usually produces jelly-covered eggs in water	Eyes but no ears; lateral line along body for detecting vibrations in water	Cold blooded; gills for breathing	Herring, rohu, shark
Amphibians	Moist skin	Four limbs; back feet often webbed to make swimming more efficient	Produces jelly-covered eggs in water	Eyes and ears	Cold blooded; lungs and skin for breathing	Frog, toad, salamander
Reptiles	Dry, with scales	Four legs (apart from snakes)	Eggs with rubbery, waterproof shell; eggs are laid on land	Eyes and ears	Cold blooded; lungs for breathing	Crocodile, python
Birds	Feathers, scales on legs	Wings; two legs	Eggs with hard shell	Eyes and ears	Warm blooded; lungs for breathing; beak	Flamingo, kestrel, pigeon
Mammals	Fur	Four limbs	Live young	Eyes, ears with pinna (external flap)	Warm blooded; lungs for breathing; females have mammary glands to produce milk to feed young; four types of teeth	Elephant, mouse

Sample question

Animals A, B and C are vertebrates:

- A has a scaly skin, four legs and lungs.
- B has hair, four legs and mammary glands.
- C has a scaly skin, fins and gills.

Create a table to show the group of organisms that each of the animals belongs to. [3]

Student's answer

Animal	Vertebrate group
A	Reptile ✓
B	Mammel ✓
C	Fish ✓

Teacher's comments

This answer has gained all three marks. The teacher allowed the second answer, although the spelling of mammal was not correct. Try to make sure that your spellings are correct – poor spelling can result in a mark not being awarded, especially if the word is similar to another biological word, for example meiosis and mitosis.

Classification of arthropods

Special features of arthropods:

- They are invertebrates – they have no backbone.
- They have an exoskeleton that is waterproof. This makes arthropods an extremely successful group, because they can exist in very dry places and are not confined to water or moist places like most other invertebrates.
- Their bodies are segmented.
- They have jointed limbs (the exoskeleton would otherwise prevent movement).

There are more arthropods than any other group of animals, so they are divided into classes. Figure 1.1 shows the differences between the four classes – insects, arachnids, crustaceans and myriapods. You only need to know about their external features.

 Insects, e.g. dragonfly, locust

Key features:
- three pairs of legs
- usually have two pairs of wings
- one pair of antennae
- body divided into head, thorax and abdomen
- a pair of compound eyes

 Arachnids, e.g. spider, tick

Key features:
- four pairs of legs
- body divided into cephalothorax and abdomen
- several pairs of simple eyes
- chelicerae for biting and poisoning prey

 Crustaceans, e.g. crab, woodlouse

Key features:
- five or more pairs of legs
- two pairs of antennae
- body divided into cephalothorax and abdomen
- exoskeleton often calcified to form a carapace (hard)
- compound eyes

 **Myriapods**, e.g. centipede, millipede

Key features:
- ten or more pairs of legs (usually one pair per segment)
- one pair of antennae
- body not obviously divided into thorax and abdomen
- simple eyes

▲ **Figure 1.1 Classification of arthropods**

Be careful with when answering questions about the different numbers of legs in insects, arachnids and crustaceans. Students often state that insects have three legs instead of three pairs of legs, losing the mark through carelessness or haste.

Five kingdoms

In the classification of living organisms, there are five kingdoms, each with its own special and obvious features. The kingdoms are as follows:

- Animals – multicellular organisms that have to obtain their food. Their cells do not have walls.
- Plants – multicellular organisms with the ability to make their own food through photosynthesis because of the presence of chlorophyll. Their cells have walls (containing cellulose).
- Fungi – many are made of hyphae, with nuclei and cell walls (containing chitin) but no chloroplasts.
- Prokaryotes (bacteria) – very small and single celled, with cell walls but no nucleus.
- Protoctists – single celled with a nucleus. Some have chloroplasts.

Features of the plant kingdom

You only need to learn the features of flowering plants and ferns.

Flowering plants (Figure 1.2) are all multicellular organisms. Their cells have cellulose cell walls and sap vacuoles. Some of the cells contain chloroplasts. They have roots, stems and leaves. Reproduction can be by producing seeds, although asexual reproduction is also possible.

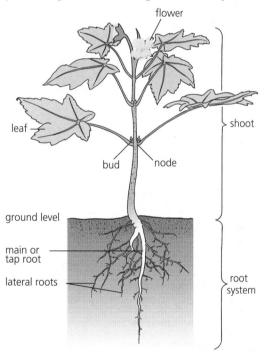

▲ **Figure 1.2 Structure of a typical flowering plant**

There are two groups – monocotyledons and dicotyledons. The term cotyledon means 'seed leaf'. The main differences between the two groups are shown in Figure 1.3 and listed in Table 1.2.

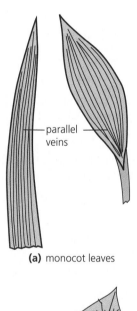

(a) monocot leaves

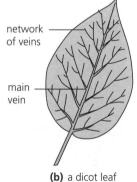

(b) a dicot leaf

▲ **Figure 1.3 Leaf types in flowering plants**

▼ Table 1.2 Features of monocotyledons and dicotyledons

Feature	Monocotyledon	Dicotyledon
Leaf shape	Long and narrow	Broad
Leaf veins	Parallel	Branching
Cotyledons	One	Two
Grouping of flower parts, such as petals, sepals and carpels	In threes	In fives

Ferns (Figure 1.4) are land plants. Their stems, leaves and roots are very similar to those of the flowering plants. The stem is usually entirely below ground. The stem and leaves have sieve tubes and water-conducting cells. Ferns also have multicellular roots with vascular tissue. The leaves are several cells thick. Most of these have an upper and lower epidermis, a layer of palisade cells and a spongy mesophyll.

Ferns do not form buds. The midrib and leaflets of the young leaf are tightly coiled and unwind as it grows. Ferns produce gametes but no seeds. The zygote gives rise to the fern plant, which then produces single-celled spores from numerous **sporangia** (spore capsules) on its leaves. The sporangia are formed on the lower side of the leaf.

Features of viruses

Viruses are very small (one-hundredth the size of bacteria), and they do not have a typical cell structure (Figure 1.5). The only life process they show is reproduction (inside host cells). They contain a strand of genetic material (DNA or RNA) and are surrounded by a protein coat.

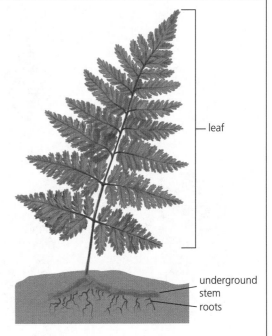

— leaf

— underground stem
— roots

▲ Figure 1.4 Structure of a fern

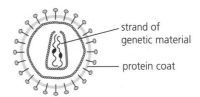

— strand of genetic material

— protein coat

▲ Figure 1.5 Structure of a virus

Revision activity

Make your own mnemonic for the five kingdoms, using the letters P, P, F, P, A.

Exam-style questions

1 Complete the following sentences about the characteristics of living organisms using only words from the list below. [4]

excretion   growth   movement   nutrition
respiration   sensitivity

A living organism can be compared to a machine such as a car. The

supply of petrol for the car is similar to _____, and the release

of energy when the petrol is burned resembles _____ in a living

organism. This can bring about the _____ of the wheels. _____ in

living organisms is similar to the release of exhaust fumes by the car.

2 Figure 1.6 shows single leaves from six different trees.

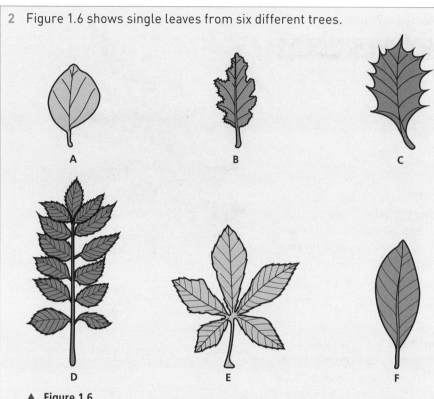

▲ Figure 1.6

Use the key below to identify which tree each leaf comes from. Make a table similar to the one below and put a tick in the correct box to show how you identified each leaf. Give the name of the tree. Leaf A has been identified for you as an example.　　　　　　　　　　　[5]

1	a	Leaf with smooth outline	go to 2
	b	Leaf with jagged outline	go to 3
2	a	Leaf about the same length as width	Cydonia
	b	Leaf about twice as long as it is wide	Magnolia
3	a	Leaf divided into more than two distinct parts	go to 4
	b	Leaf not divided into more than two distinct parts	go to 5
4	a	Leaf divided into five parts	Aesculus
	b	Leaf divided into ten or more parts	Fraxinus
5	a	Leaf with pointed spines along its edge	Ilex
	b	Leaf with rounded lobes along its edge	Quercus

Leaf	1a	1b	2a	2b	3a	3b	4a	4b	5a	5b	Name of tree
A	✓		✓								Cydonia
B											

3 Figure 1.7 shows some invertebrates found in a compost heap.

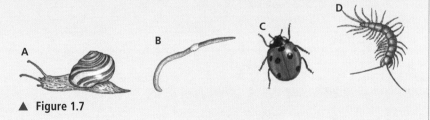

▲ Figure 1.7

Use the key to identify each animal and state the items in the key used in each identification. One has been done for you. [6]

1 Has legs   2
    No legs   5
2 More than six legs   3
    Six legs   4
3 Short, flattened grey body   *Oniscus asellus*
    Long brown/yellow body   *Lithobius forficatus*
4 Pincers on last segment   *Forficula auricularia*
    Hard wing covers   *Coccinella septempunctata*
5 Body segmented   *Lumbricus terrestris*
    Body not segmented   6
6 Has a shell   *Helix aspersa*
    No shell   *Arion ater*

Animal	Name of animal	Items used in the key
A	*Helix aspersa*	1, 5, 6
B		
C		
D		

4 Figure 1.8 can be used to identify the main classes of vertebrate. Use the key to identify the main classes represented by the letters A–E. [5]

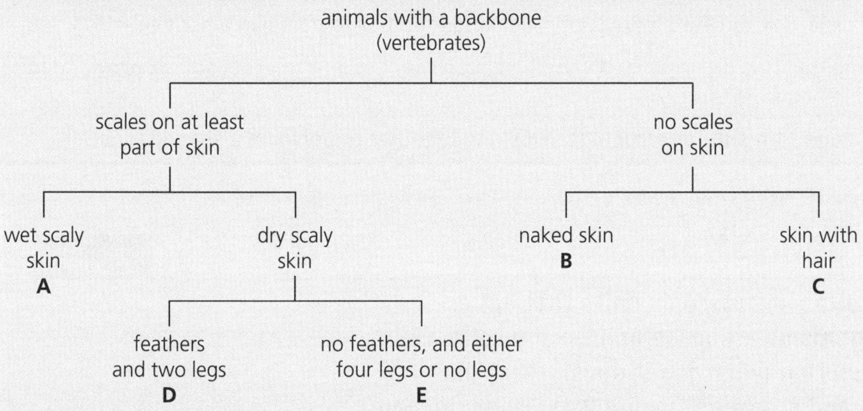

▲ **Figure 1.8**

5 a Copy the diagrams of the insect, crustacean and arachnid in Figure 1.1 (p. 5) and label the key features that you can see. [4]
   b Copy the myriapod diagram in Figure 1.1 and label the features that are common to all arthropods. [3]

Key objectives

The objectives for this chapter are to revise:
● definitions of the key terms
● structures of plant, animal and bacterial cells and the functions of cell structures
● tissues, organs and organ systems
● calculating the magnification and size of biological specimens

● how to convert measurements between millimetres (mm) and micrometres (μm)

Key terms

REVISED

Term	Definition
Cell	The smallest basic unit of an animal or plant; it is microscopic and acts as a building block
Magnification	The observed size of an image divided by the actual size of the image
Organ	A structure made up of a group of tissues working together to perform a specific function
Organ system	A group of organs with related functions working together to perform a body function
Organism	A living thing that has an organised structure, can react to stimuli, reproduce, grow, adapt, and maintain homeostasis
Tissue	A group of cells with similar structures working together to perform a shared function

Cell structure and organisation

REVISED

Most living things are made of **cells** – microscopic units that act as building blocks. Multicellular **organisms** are made up of many cells. Cell shape varies depending on its function (what job it does). Plant and animal cells differ in size, shape and structure (Figure 2.1). Plant cells are usually larger than animal cells.

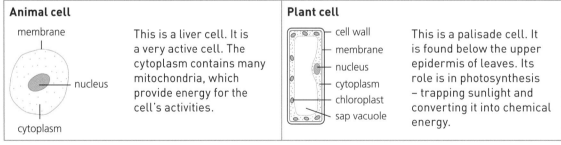

Animal cell

membrane
nucleus
cytoplasm

This is a liver cell. It is a very active cell. The cytoplasm contains many mitochondria, which provide energy for the cell's activities.

Plant cell

cell wall
membrane
nucleus
cytoplasm
chloroplast
sap vacuole

This is a palisade cell. It is found below the upper epidermis of leaves. Its role is in photosynthesis – trapping sunlight and converting it into chemical energy.

▲ Figure 2.1 Comparison of animal and plant cells

When viewed under an electron microscope, other organelles become visible. These include ribosomes and mitochondria (Figure 2.2).

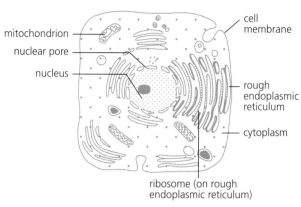

Labels: mitochondrion, nuclear pore, nucleus, cell membrane, rough endoplasmic reticulum, cytoplasm, ribosome (on rough endoplasmic reticulum)

▲ **Figure 2.2 Liver cell**

Remember:

- Animal cells contain only three main parts: membrane, nucleus and cytoplasm.
- Animal cells never have a cell wall, chloroplasts or sap vacuoles (although they may have temporary vacuoles where food is stored).
- Not all cells have all cell parts when mature – for example, red blood cells do not have a nucleus and xylem cells do not have a nucleus or cytoplasm.
- It is not true that all plant cells contain chloroplasts – for example, epidermis cells and root cells do not.
- Chloroplasts (structures or organelles) are different from chlorophyll (the chemical found in them).

> **Revision activity**
>
> Make a mnemonic to help you remember the three main parts of animal cells (membrane, nucleus, cytoplasm) – for example, Mice Nibble Cheese.

Parts of a cell

Structures found in animal and plant cells are summarised in Table 2.1.

▼ **Table 2.1 Structures in animal and plant cells, and their functions**

	Part	Description	Where found	Function
Animal and plant cells	Cytoplasm	Jelly-like, containing particles and organelles	Enclosed by a cell membrane	Contains cell organelles, e.g. mitochondria, nucleus Chemical reactions take place here
	Membrane	Partially permeable layer that forms a boundary around the cytoplasm	Around the cytoplasm	Prevents cell contents from escaping Controls what substances enter and leave the cell
	Nucleus	Round or oval structure containing DNA in the form of chromosomes	Inside the cytoplasm	Controls cell division Controls cell development Controls cell activities
	Ribosomes	Tiny particles floating freely or attached to membranes called rough endoplasmic reticulum	Inside the cytoplasm	Responsible for synthesis of proteins from amino acids
	Mitochondria (singular: mitochondrion)	Circular, oval or slipper-shaped organelle	Inside the cytoplasm	Responsible for aerobic respiration Cells with high rates of metabolism, e.g. liver cells, require large numbers of mitochondria to provide sufficient energy

	Part	Description	Where found	Function
Plant cells only	Cell wall	Tough, non-living layer made of cellulose; it surrounds the membrane	Around the outside of plant cells	Prevents plant cells from bursting Freely permeable (allows water and mineral ions to pass through)
	Sap vacuole	Fluid-filled space surrounded by a membrane	Inside the cytoplasm of plant cells	Contains mineral ions and sugars Helps keep plant cells firm
	Chloroplasts	Organelles containing chlorophyll	Inside the cytoplasm of some plant cells	Trap light energy for photosynthesis

Sample question
REVISED

Figure 2.3 shows a nerve cell. State the names of the cell parts A, B and C. [3]

▲ Figure 2.3

Student's answer

A: cell wall ✗; B: cytoplasm ✓; C: nucleus ✓

Revision activity

Trace, copy or sketch the cells shown in Figure 2.1 (p. 10). Practise labelling both cells. Then do the same with other types of animal and plant cells.

When labelling plant cells, start from the outside (the cell wall) and work inwards on this order: cell wall, membrane, cytoplasm, chloroplast, nucleus, sap vacuole. The chloroplasts and nucleus are both held inside the cytoplasm.

Cell walls are always drawn as a double line to show their thickness. Make sure that your cell wall label line touches the outer line. The membrane label line should touch the inner line of the cell wall (when plant cells are turgid – firm – the membrane is pressed against the cell wall).

Teacher's comments

The first answer is wrong – a nerve cell is an animal cell, so it does not have a cell wall. The correct answer for part A is cell membrane.

Bacterial cell structure

Bacteria are very small organisms that are single cells. They have a cell wall surrounding the cytoplasm, which contains large numbers of free-floating ribosomes and granules. Bacteria do not have a nucleus: they have circular DNA, made of a single chromosome, and plasmids (Figure 2.4).

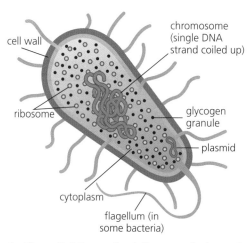

cell wall

chromosome
(single DNA
strand coiled up)

ribosome

glycogen
granule

plasmid

cytoplasm

flagellum (in
some bacteria)

▲ **Figure 2.4 Generalised diagram of a bacterium**

▼ **Table 2.2 Structures in bacterial cells and their functions**

Part	Description	Where found	Function
Cytoplasm	Jelly-like; contains particles and organelles	Surrounded by the cell membrane	Contains cell structures, e.g. ribosomes, circular DNA, plasmids
Cell membrane	A partially permeable layer that surrounds the cytoplasm	Around the cytoplasm	Prevents cell contents from escaping Controls what substances enter and leave the cell
Circular DNA	A single circular chromosome	Inside the cytoplasm	Controls cell division Controls cell development Controls cell activities
Plasmids	Small, circular pieces of DNA	Inside the cytoplasm	Contain genes that carry genetic information to help the processes of survival and reproduction of the bacterium
Ribosomes	Small, circular structures	Inside the cytoplasm	Protein synthesis
Cell wall	A tough, non-living layer (not made of cellulose) that surrounds the cell membrane	Around the outside of the bacterial cell	Prevents the cell from bursting Allows water and mineral ions to pass through (freely permeable)

Formation of new cells

Cells have a limited lifespan; if they become damaged, they may not function properly. New cells are constantly being formed by the division of existing cells.

Plant and animal cells divide by a process called **mitosis** (Chapter 17). Gametes (sex cells) are formed by a different process called **meiosis**, which involves halving the chromosome number (Chapter 17).

Bacterial cells do not divide by mitosis because they do not have a nucleus. They divide in the process of asexual reproduction (Chapter 16).

Specialisation of cells

Figure 2.5 shows examples of cells and their functions in **tissues**.

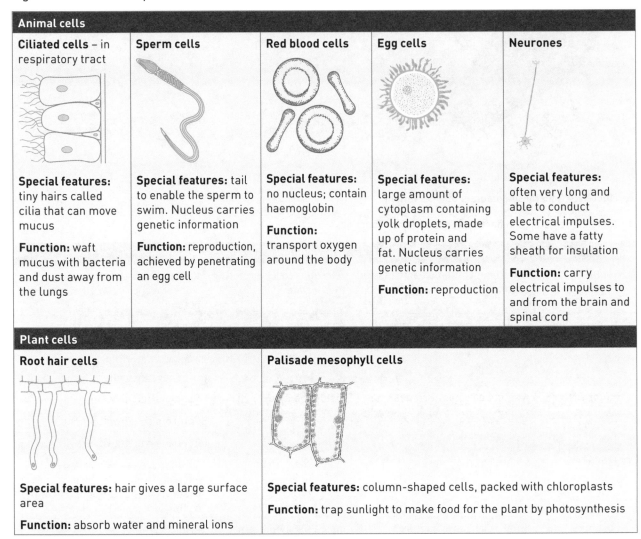

▲ **Figure 2.5 Examples of specialised animal and plant cells**

Xylem and phloem tissues are often confused. Remember:

- Xylem carries water and mineral ions.
- Phloem transports sugars and amino acids.
- In a vascular bundle in a stem, phloem is on the outside and xylem is on the inside.

Sample question

With reference to a suitable named example, define the term *tissue*.   [3]

Student's answer

A tissue is a group of cells ✓ carrying out the same job ✓.

Teacher's comments

The answer needs three clear points to gain the 3 marks available. This student has not named a type of tissue (even though this was the first instruction in the question) and has given only two correct points. Always use the marks shown in the margin to show you how many points to give. Avoid giving more than three; this would waste time that you might need to answer other questions. Choose three statements to make before writing them down. The teacher will not select the best answers from a mixture of good and bad ones.

Organs and organ systems

REVISED

You need to be able to give examples of **organs** and **organ systems** in both plants and animals.

Organs are made of several tissues grouped together to make a structure with a special job. A leaf is an organ made up of a number of tissues, such as the upper epidermis and palisade mesophyll. Organ systems are groups of organs with closely related functions. Table 2.3 shows examples found in animals and plants.

▼ Table 2.3 Examples of organs and organ systems in animals and plants

Organism	Examples of organs	Examples of organ systems
Animal	Heart, lungs, intestine, eye, brain	Circulatory system, nervous system, digestive system
Plant	Leaf, stem, flower	Shoot, reproductive system

Revision activity

It is important that you can identify the different levels of organisation in drawings, diagrams and images of plant and animal material. Practise this by looking at examples in textbooks or on the internet.

Annotating (adding a description to a labelled part of a diagram or drawing) these diagrams and drawings is also a useful revision tool, and may help you gain extra marks in an exam answer. Figure 2.6 shows the action of a phagocyte, for example.

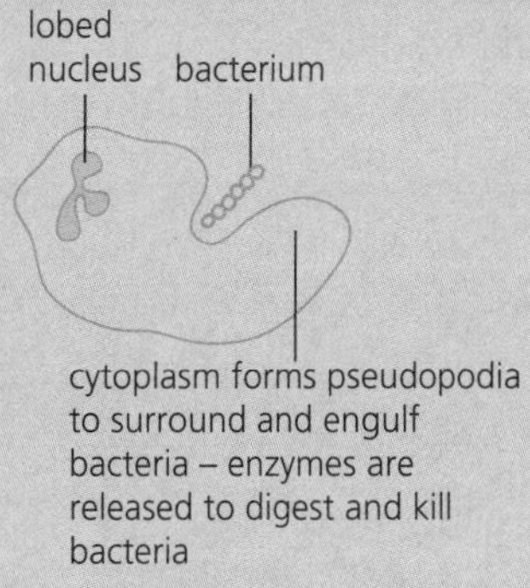

lobed nucleus   bacterium

cytoplasm forms pseudopodia to surround and engulf bacteria – enzymes are released to digest and kill bacteria

▲ **Figure 2.6 A phagocyte engulfing a bacterium**

Size of specimens

REVISED

A microscope makes a specimen appear larger than it really is (it magnifies the specimen). You need to be able to calculate the **magnification** and also the actual size of the specimen.

If dealing with a very large number, it may be better to display it in standard form.

Skills

Standard form

Standard form is a way of writing down very large or very small numbers more easily. It uses the powers of 10 to show how big or how small the number is.

You write it as $y \times 10^n$ where:
- y is always a number greater than or equal to 1, but less than 10
- n can be any positive or negative whole number

If answering an extended paper, remember that there are 1000 micrometres (μm) in a millimetre. Therefore:

- To change a measurement from micrometres to millimetres, you need to divide the figure by 1000.
- To change a measurement from millimetres to micrometres, you need to multiply the figure by 1000.

Skills

Magnification

To calculate the magnification of specimens that have been observed using a light microscope, memorise and use the following equation:

$$\text{magnification} = \frac{\text{observed size}}{\text{actual size}}$$

Make sure that the observed size and actual size have the same units.

Exam-style questions

1   a   Describe how a bacterial cell is different from a plant cell such as a palisade cell. [3]
    b   Explain why bacterial cells do not divide by mitosis. [1]
2   Identify parts A, B, C and D shown in Figure 2.7, and describe their main features and functions. [12]

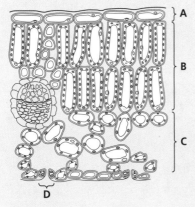

▲ Figure 2.7

3   a   Name one organ *not* given in Table 2.3 that is found in:
        i   animals [1]
        ii plants [1]
    b   Name two tissues found in each of the organs you have named. [4]

4   The diagram of a cow's eye shown in Figure 2.8 is magnified ×2.5 (not drawn to scale). Calculate the actual width of the eye, as shown between points A and B. Show your working. [2]

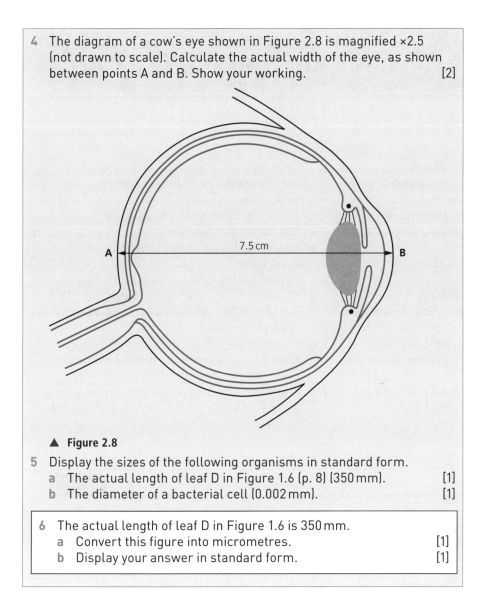

7.5 cm

A ◄———————————————► B

▲ **Figure 2.8**

5   Display the sizes of the following organisms in standard form.
   a   The actual length of leaf D in Figure 1.6 (p. 8) (350 mm). [1]
   b   The diameter of a bacterial cell (0.002 mm). [1]

6   The actual length of leaf D in Figure 1.6 is 350 mm.
   a   Convert this figure into micrometres. [1]
   b   Display your answer in standard form. [1]

3 Movement into and out of cells

Key objectives

The objectives for this chapter are to revise:
- definitions of diffusion and active transport
- the source of energy for diffusion
- that some substances move into and out of cells by diffusion through the cell membrane
- the importance of diffusion of gases and solutes in living organisms
- the factors that influence diffusion
- the role of water as a solvent in organisms
- that water diffuses through partially permeable membranes, and into and out of cells through the cell membrane, by osmosis
- investigations into the effects on plant tissues of immersing them in solutions of different concentrations

- that plants are supported by the pressure of water inside the cells pressing outwards on the cell wall

- the definition of osmosis
- how to explain the effects of osmosis on plant cells
- how to use the terms associated with osmosis
- the importance of water potential and osmosis in the uptake and loss of water by organisms
- how to explain the importance of active transport as a process for movement of molecules or ions across membranes
- that protein carriers move molecules or ions across a membrane during active transport

Key terms

REVISED

Term	Definition
Active transport	The movement of particles through a cell membrane from a region of their lower concentration to a region of higher concentration (i.e. against a concentration gradient), using energy from respiration
Diffusion	The net movement of particles from a region of their higher concentration to a region of lower concentration (i.e. down a concentration gradient), as a result of their random movement
Osmosis	The net movement of water molecules from a region of higher water potential (dilute solution) to a region of lower water potential (concentrated solution) through a partially permeable membrane

Diffusion

REVISED

Diffusion is a really important process for living organisms because it helps to provide essential gases and solutes (materials in solution), and also helps to remove some substances that are potentially toxic (poisonous). These move into or out of the cell through the cell membrane. Table 3.1 gives some examples.

▼ Table 3.1 Examples of diffusion in living organisms

Site of diffusion	Substance	Description
Alveoli of lungs	Oxygen	From the alveoli into the blood capillaries
Alveoli of lungs	Carbon dioxide	From blood capillaries into the alveoli
Stomata of leaf	Oxygen	From the air spaces, through stomata, into the atmosphere during photosynthesis

The energy for diffusion comes from the kinetic (movement) energy of the random movement of molecules and ions. From the organism's point of view, it is a 'free' process – no energy needs to be provided to make it happen.

Rates of diffusion

You need to be able to state the factors that help diffusion to be efficient. These are:

- distance (the shorter the better), for example the thin walls of alveoli and capillaries
- concentration gradient (the bigger the better); this can be maintained by removing the substance as it passes across the diffusion surface (think about oxygenated blood being carried away from the surface of alveoli)
- surface area for diffusion (the larger the better), for example there are millions of alveoli in a lung, giving a huge surface area for the diffusion of oxygen
- temperature (molecules have more kinetic energy at higher temperatures)

Do not confuse cell walls with capillary walls – animal cells do not have walls. Many students throw away marks by referring to 'the thin cell walls of capillaries'. What they mean is 'the *walls of capillaries* are thin because they are only one cell thick'.

Skills

Investigating how distance travelled affects the rate of diffusion

You may be asked how to investigate how the distance a material has to travel affects the rate of diffusion. This can be done using transparent blocks of agar or gelatine. They are cut into four cubes, all with length of side 3.0 cm, and placed in a beaker containing a dye, such as methylene blue.

After 15 minutes, the first cube is removed and cut in half. The depth to which the dye has diffused is measured. The other cubes are removed at intervals of 15 minutes, sectioned and measured in the same way. The rate of diffusion for each cube is calculated using the equation:

$$\text{rate} = \frac{\text{distance travelled}}{\text{time}}$$

Your results should lead you to the following conclusions:

1. The longer the cube is left in the dye, the greater the distance travelled by the dye.
2. The further the dye travels, the slower the rate of diffusion.

Osmosis

REVISED

Water is important to living things as a solvent – many substances (solutes) dissolve in it. Examples include glucose, mineral ions and amino acids. In animals, water is essential for the following processes:

- **Digestion** – water helps to break down and dissolve food molecules in the process of digestion.
- **Transport** – blood is made up of cells and a water-based liquid called plasma. The plasma is a way of transporting many dissolved substances, for example carbon dioxide, urea, glucose and hormones.
- **Excretion** – water is important in the process of excretion in animals because some of the excretory materials, for example urea, are toxic. Water dilutes these to make them less poisonous. Urine is a solution containing dissolved mineral ions, urea, used hormones and drugs.

Osmosis is a special form of diffusion. It always involves the movement of water across a partially permeable membrane. Plants rely on osmosis to obtain water through their roots. They use water as a transport medium to carry dissolved mineral ions, sucrose and amino acids around the plant through the xylem and phloem vessels, and to maintain the firmness of cells. When young plants lose more water than they gain, cells become limp and the plants wilt.

Fish living in salt water lose water by osmosis. They have very efficient kidneys to reduce water loss in urine.

If we get dehydrated, water is lost from our red blood cells by osmosis. As the cells shrink, they become less efficient at carrying oxygen.

Effects of osmosis on plant and animal tissues

- When placed in water, plant and animal cells will take in the water through their cell membranes by diffusion. The diffusion of water in this way is called osmosis.
- Plant cells become swollen, but do not burst because of their tough cell wall.
- Plants are supported by the pressure of water inside the cells pressing outwards on the cell wall, which is inelastic and prevents further net entry of water.
- Animal cells will burst because they have no cell wall.
- The reverse happens when plant and animal cells are placed in concentrated sugar or salt solutions – plant and animal cells become limp.

Remember that sugars and mineral ions do *not* move by osmosis. Cell membranes can prevent some substances entering or leaving the cell.

Sample question

REVISED

Some sugar solution was collected from the phloem of a plant stem. Plant cells were placed on a microscope slide and covered with this sugar solution.

Describe what changes would occur to each of the following three cell parts if the sugar solution was more concentrated than the sap in the cell vacuole: sap vacuole, cytoplasm, cell wall.　　　　[3]

Student's answer

Sap vacuole: this will get smaller ✓ because there is a higher concentration of water inside the cell, so the water will pass out of the vacuole by osmosis.

Cytoplasm: this will shrink because it is losing water. ✗

Cell wall: this will stop stretching and stop curving outwards. ✓

Correct answer

The sap vacuole will get smaller.

The cytoplasm will shrink and pull away from the cell wall.

The cell wall will stop stretching and stop curving outwards.

Teacher's comments

The first answer is correct, but this student has wasted time writing more than is needed – the question required a description, not an explanation. The second answer should give details about the way the cytoplasm comes away from the cell wall. In the third answer, details about the cell wall are not very well worded, but it is clear that the student understands what is happening.

Skills

Investigating osmosis with a partially permeable membrane

Materials that act as partially permeable membranes, such as dialysis tubing, can be used to investigate osmosis. The tubing allows small molecules and ions to pass through but prevents the movement of larger molecules. You need to be aware of how you can investigate osmosis using a material such as dialysis tubing. Figure 3.1 shows how dialysis tubing can be used in an osmosis investigation.

If the apparatus is left for 30 minutes, the dialysis tubing will swell. The pressure forces the coloured solution up the capillary tube and may even run out of the top. This happens because water enters the dialysis tubing by osmosis. The increase in volume of liquid increases the pressure, forcing the solution up the capillary tube.

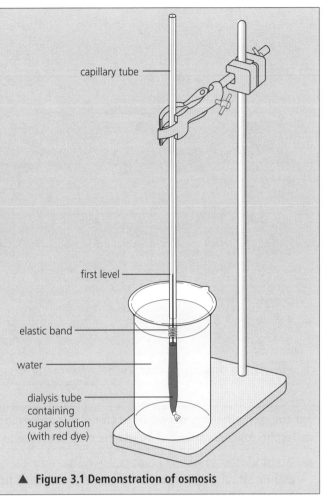

capillary tube

first level

elastic band

water

dialysis tube containing sugar solution (with red dye)

▲ **Figure 3.1 Demonstration of osmosis**

Skills

Investigating the effect of solutions of different concentrations on plant tissues

You need to be able to describe investigations of the effects on plant tissues of immersing them in solutions of different concentrations.

To do this, all the variables in the experiment, except the factor under test (in this case, the concentration of solution), need to be kept the same. The variables can include the size of material (e.g. potato cylinders), type of material (same variety of potato), volume of solution, time the material is left in the solutions, temperature, and size of containers. The measuring instruments (ruler or top-pan balance and measuring cylinder for the solutions) also need to be the same for all measurements. Each of the containers needs to be labelled to identify the solution in it.

A range of concentrations of solution (which may be sugar or salt) is set up. Pieces of the plant material are cut to the same size. Cutting the potato into cylinders is helpful because they will all have the same diameter; they can then be cut to the same length, or weighed. The potato pieces are left in the solutions for a fixed amount of time before being removed and remeasured (Figure 3.2).

Plant tissues become longer and gain mass in very weak solutions because the cells take in water by osmosis. Those in more concentrated solutions become shorter and lose mass because the cells lose water by osmosis. There is a concentration where there appears to be no change in length or mass because the concentration of the solution is equivalent to the concentration of the sap in the cells of the plant tissue.

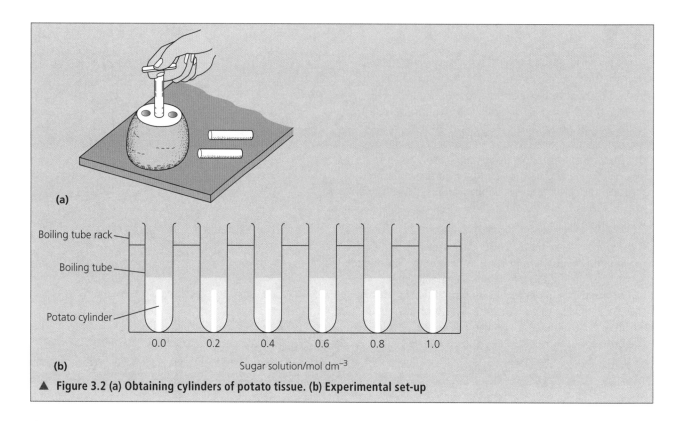

(a)

Boiling tube rack

Boiling tube

Potato cylinder

0.0   0.2   0.4   0.6   0.8   1.0

(b)   Sugar solution/mol dm^{-3}

▲ **Figure 3.2 (a) Obtaining cylinders of potato tissue. (b) Experimental set-up**

For the extended exam, you need to be able to explain the effects of different concentrations of solutions on plant tissues:

- When placed in water, plant cells will take in the water through their cell membranes because there is a higher **water potential** outside the cells than inside.
- The term *water potential* means the potential for water to move.
- Water always moves from a higher water potential to a lower water potential.
- A weak (dilute) solution (or pure water) has a high water potential; a strong (concentrated) solution has a low water potential.
- Plant cells become **turgid** (swollen), but do not burst because of their tough cell wall. A **turgor pressure** is created, which will restrict any further entry of water into the cell.
- The reverse happens when plant cells are placed in concentrated sugar or salt solutions. This is because there is a higher water potential inside the cell than outside it.
- Plant cells become **flaccid** (limp) and the cytoplasm is no longer pressed against the cell wall. The process of water loss from a cell when it is placed in a solution with a lower water potential is called **plasmolysis**.

Water potential and osmosis are important in the uptake and loss of water by organisms. If all the cells in a plant's leaves and stem are turgid the plant will remain upright and the leaves will be held out straight to capture sunlight. This is achieved by the cells taking in water by osmosis. This can only happen if the water potential in the plant cells is lower than that of the liquid surrounding them.

If root hair cells are bathed by a solution with a lower water potential than that of the cell sap (for example, if concentrated chemical fertiliser has been added to the soil) water will leave the cells by osmosis. This can kill the plant. Plants that have lost water by osmosis due to the difference in water potential between cells (higher water potential) and the surrounding fluid cells (lower water potential) are described as 'wilting'.

It is important that the fluid that bathes animal cells, like tissue fluid or blood plasma, has the same water potential as the cells' contents:

- If the bathing fluid has a higher water potential, water will enter the cells by osmosis. The cells will swell and burst, because they have no cell wall.
- If the bathing fluid has a lower water potential, water will leave the cells by osmosis. The cells will become plasmolysed and may lose their function. This can happen to red blood cells, for example if an athlete becomes dehydrated. When the cells become plasmolysed, their surface area is reduced and they can absorb and carry less oxygen.

Skills

Calculating percentage change

One of the mathematical skills you may be asked to demonstrate is the ability to calculate percentage change. Osmosis investigations can provide an opportunity to show this skill.

The formula for percentage change is:

$$\frac{\text{change in size}}{\text{size at start}} \times 100$$

For example, if a potato cylinder is 6.0 cm long and increases in length to 6.5 cm after being placed in water, its increase in length is: 6.5 − 6.0 = 0.5 cm.

Applying the formula, the percentage change in length is:

$$\frac{0.5}{6.0} \times 100 = 8.3\%$$

Active transport

REVISED

Active transport is the movement of particles through a cell membrane from a region of lower concentration to a region of higher concentration (i.e. against a concentration gradient) using energy from respiration.

Note the two big differences between diffusion and active transport:

- The direction of movement (down a gradient, or up a gradient).
- Whether or not energy is needed for the movement.

Animals and plants rely on active transport to move some substances because the concentration gradient is not always the right way round for diffusion. However, cells need to provide energy to achieve movement by active transport. This energy is supplied through respiration. Mitochondria (cell organelles in the cytoplasm) control energy release. Respiratory poisons block energy release, so they can prevent active transport.

Protein molecules, called carriers, in the cell membrane play an important part in moving particles across the membrane. The protein uses energy to move the particles against their concentration gradient.

Table 3.2 gives some examples of active transport.

▼ Table 3.2 Examples of active transport in living organisms

Site of active transport	Substance	Description
Root hair cells	Mineral ions, such as phosphate	From the soil into the roots
Wall of small intestine (villi)	Glucose	From the small intestine into the blood plasma
Kidney tubules	Glucose	From the filtrate in the tubule into a blood capillary

Sample question

REVISED

Materials can enter cells by diffusion, osmosis and active transport.
a Define the term *diffusion*. [3]
b Describe how the following processes differ from diffusion:
  i osmosis [2]
  ii active transport [2]

Student's answer

a Movement of particles ✗ from a region of higher concentration to a region of lower concentration ✓.
b i Osmosis only involves the movement of water. ✓ With osmosis, water passes through the cell wall. ✗
  ii Active transport requires energy from respiration ✓ and moves particles against the concentration gradient ✓.

Correct answer

a The net movement of particles from a region of higher concentration to a region of lower concentration due to the random movement of the particles.
b i Osmosis only involves the movement of water. With osmosis, water passes through a partially permeable membrane.
  ii Active transport requires energy from respiration and moves particles against the concentration gradient.

Teacher's comments

In part a, the student has gained only one mark from the three available. The first part of the definition needed to be more accurate, by referring to the *net* movement of particles. The statement about the concentration gradient is correct, but the student has forgotten to state that the movement is a result of the random movement of the particles, which was the third marking point.

In b(i), the second statement has not been awarded a mark because both processes may involve movement of particles through a cell wall (if the cell is a plant cell), but osmosis involves movement of water through a partially permeable membrane.

In b(ii) the student has given a very good answer, gaining both available marks.

Exam-style questions

1 A potato was set up as shown in Figure 3.3 (left-hand side). The investigation was left for several hours. The results are shown on the right-hand side of the figure.

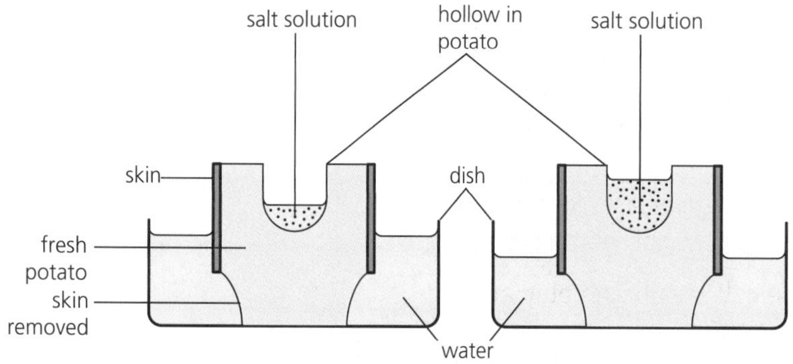

▲ Figure 3.3

a   Describe what happened to:
    i   the water in the dish   [1]
    ii the salt solution in the hollow in the potato   [1]
b   i   Name the process that is responsible for the changes
        that have occurred.   [1]
    ii Explain why these changes have occurred.   [3]
    iii Where does this process occur in a plant?   [1]
    iv What is the importance to the plant of this process?   [1]

2 A cylinder of potato of mass 20.0 g was placed in concentrated
   sugar solution. When reweighed, it had a mass of 18.5 g.
   a   Calculate the percentage change in mass of the potato
       cylinder.   [2]
   b   Explain why the potato cylinder lost mass.   [3]

3 Figure 3.4 shows part of a root.

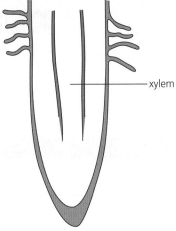

—xylem

▲ **Figure 3.4**

a   Explain how the presence of root hair cells on roots enables
    the efficient absorption of water and mineral ions.   [2]
a   Root hair cells absorb mineral ions by active transport.
    i   Define the term *active transport*.   [2]
    ii Explain why respiration rates may increase in root hair
        cells during the uptake of mineral ions.   [1]

4 Complete the table by stating the process involved in the
   movement of the material in each description.   [3]

Description	Process
Uptake of water in a plant by root hairs	
Loss of water vapour from the stomata of a leaf	
Uptake of mineral ions into a root by protein carriers	

Key objectives

The objectives for this chapter are to revise:
- the chemical elements that make up carbohydrates, fats and proteins
- the synthesis of large molecules from smaller base units
- food tests for starch, reducing sugars, proteins, fats and vitamin C
- the structure of a DNA molecule

Biological molecules

REVISED ☐

Carbohydrates, fats and proteins are all made up of the elements carbon, oxygen and hydrogen. In addition, proteins always contain nitrogen; they sometimes also contain sulfur or phosphorus. Table 4.1 summarises the main nutrients.

▼ **Table 4.1 Summary of the main nutrients**

Nutrient	Elements present	Sub-units	Examples
Carbohydrate	Carbon, hydrogen, oxygen	Glucose	Starch, glycogen, cellulose, sucrose
Fat/oil (oils are liquid at room temperature but fats are solid)	Carbon, hydrogen, oxygen (but lower oxygen content than carbohydrate)	Fatty acids and glycerol	Vegetable oils, e.g. olive oil Animal fats, e.g. cod liver oil, waxes
Protein	Carbon, hydrogen, oxygen, nitrogen; sometimes also sulfur or phosphorus	Amino acids (about 20 different forms)	Enzymes, muscle, haemoglobin

Revision activity

One way of remembering the elements in carbohydrate is to look at its name: **C**arb **O** **H**ydrate (carbon, oxygen, hydrogen).

Large carbohydrate molecules such as starch, glycogen and cellulose are made up of long chains of smaller units – monosaccharides, such as glucose – held together by chemical bonds (Figure 4.1).

Fats and oils are made up of three units of fatty acids chemically bonded to one glycerol unit (Figure 4.2).

Proteins are made of long chains of amino acids chemically bonded together (Figure 4.3). As there are about 20 different amino acids, their pattern in the chain can be quite complex, and the molecules can be very large.

▲ **Figure 4.1 Carbohydrate**

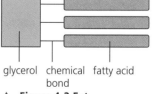

▲ **Figure 4.2 Fat**

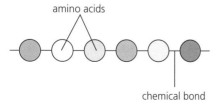

▲ **Figure 4.3 Protein**

Food tests

You need to be able to describe the tests for starch, reducing sugars, proteins and fats (Table 4.2). Make sure you also learn the colour change for a positive result.

▼ **Table 4.2 Summary of food tests**

Food tested	Name of test	Method	Positive result
Starch	starch test	Add a few drops of iodine solution to a solution of the food	Blue/black colouration
Reducing sugar	Benedict's test	Add an equal amount of Benedict's solution to a solution of the food Boil carefully	A succession of colour changes: from turquoise to pale green, pea green, orange then brick red The further the colour change is along the gradient, the more reducing sugar is present
Protein	biuret test	Add an equal amount of sodium hydroxide to a solution of the food Mix carefully Then add a few drops of 1% copper sulfate, without shaking the mixture	Violet halo
Fats	emulsion test	Dissolve the food in ethanol Pour the solution into a clean test tube of water	White emulsion
Vitamin C	DCPIP test	DCPIP is a deep blue colour Measure $2.0\,cm^3$ DCPIP solution into a test tube Add fruit juice drop by drop, counting the drops used Shake the mixture after each drop	DCPIP decolourises

Note that food tests do not usually involve heating; the only food test that does is Benedict's test.

Skills

Describing food tests

Questions on food tests often appear on the Practical and Alternative to practical examination papers. Make sure you learn each of the food tests. When you are describing them:

- include the names of the reagents needed
- include the volumes of the reagents required
- state if boiling of the mixture is required (heating is only needed for Benedict's test)
- state the colour of the reagent before and after the test
- emphasise any safety precautions needed

Sample question

REVISED

Describe how you could compare the vitamin C contents of a range of fruits. [5]

Student's answer

I would use the DCPIP test. ✓ First, I would get the juice from each of the fruits and measure out equal amounts. Add the juice to the DCPIP drop by drop ✓ and record how many drops are needed to make the DCPIP change colour. Repeat for the other juices.

Teacher's comments

This answer could be improved by including more details, such as:
- The mixture should be shaken after each drop of DCPIP has been added.
- The colour change should be stated.
- A statement that the fewest drops needed indicates the juice containing the highest concentration of vitamin C.
- A statement about the need to clean out the glassware or pipette after each test.

Correct answer

I would use the DCPIP test. First, I would juice the fruits, then I would measure 2 cm³ DCPIP solution into a test tube. Using a pipette, I would add one of the juices to the DCPIP drop by drop, shaking the mixture after each drop, and record how many drops are needed to make the DCPIP change from blue to colourless. I would repeat the test for the other juices, washing the test tube and pipette thoroughly between each test. The juice that causes the colour change with the fewest drops has the highest concentration of vitamin C.

Structure of DNA

REVISED

A DNA molecule is made up of two strands. The double strand is twisted to form a double helix (a double spring). Each strand contains chemicals called bases – A, T, C or G (Figure 4.4). Each base in one strand is cross-linked to a base in the other strand by a bond. A always bonds with T; C always bonds with G.

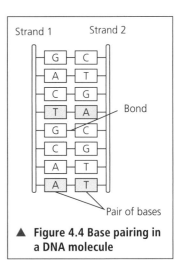

▲ Figure 4.4 Base pairing in a DNA molecule

Exam-style questions

1 Make a table for carbohydrates, fats and proteins, with the headings shown below. [9]

Biological molecule	Chemical elements present	Sub-unit(s)	Examples
Carbohydrate			
Fat/oil			
Protein			

2 A student was a testing a sample of food for the presence of protein and reducing sugar.
  a State the name of the reagent needed to test for:
    i protein [1]
    ii reducing sugar [1]
  b Which test would involve boiling the mixture? [1]

3 a Part of a DNA molecule has the base sequence
    A T T C G A C A G on one strand.
    What will be the sequence of bases on the other strand? [1]
  b i State the name of the structure formed when two strands of DNA coil together. [1]
    ii Describe how the strands of DNA are held together. [2]

5 Enzymes

Key objectives

The objectives for this chapter are to revise:
- definitions of the key terms
- why enzymes are important in all living organisms
- enzyme action
- the effects of changes in temperature and pH on enzyme activity
- how to explain enzyme action with reference to the active site, enzyme–substrate complex, substrate and product
- how to explain the specificity of enzymes
- how to explain the effect of changes in temperature and pH on enzyme activity

Key terms

REVISED

Term	Definition
Catalyst	A substance that increases the rate of a chemical reaction and is not changed by the reaction
Enzyme	Proteins that function as biological catalysts and are involved in all metabolic reactions

Enzyme action

REVISED

Most chemical reactions happening in living things are helped by **enzymes**, which act as **catalysts**. The speed at which they can catalyse reactions is sufficient to sustain life.

Most enzyme names end in -ase, for example lipase and protease. Enzymes usually speed up reactions, but some slow them down. Some enzymes control reactions to build up molecules (synthesise them), for example starch phosphorylase:

$$\text{maltose} \xrightarrow{\text{starch phosphorylase}} \text{starch}$$

Others are involved in breaking down molecules, e.g. protease in digestion:

$$\text{protein} \xrightarrow{\text{protease}} \text{amino acids}$$

(See Chapter 7 for other examples of enzymes.)

Enzyme molecules are **proteins**. Each molecule has an **active site**, which is a special shape that the substrate fits into (a **complementary shape**). Once the enzyme molecule and substrate come into contact, one or more **products** are formed.

> **Revision activity**
>
> Learn the definition of an enzyme by heart. Do this by reading it, then covering it up and writing it out. Check it is correct. Repeat this three or four times. Test yourself again 24 hours later.

Enzymes and temperature

The optimum (best) temperature for enzyme-controlled reactions is around 37°C (body temperature). Increasing the temperature above the optimum temperature slows the reaction down because the enzyme molecules become denatured (Figure 5.1).

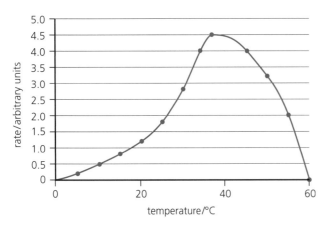

▲ **Figure 5.1 The effect of temperature on the rate of an enzyme-controlled reaction**

Enzymes and pH

The pH of a solution is how acidic or alkaline it is. Most enzymes have an optimum pH (at which they work best) – usually around neutral (pH 7) – but there are some exceptions:

- protease, pH 2.0 – in the stomach, with hydrochloric acid
- salivary amylase, pH 6.8 – in the mouth
- catalase, pH 7.6 – in plants, e.g. potato
- pancreatic lipase, pH 9.0 – in the duodenum

The 'wrong' pH slows down enzyme activity, but this can usually be reversed if the optimum pH is restored. However, an extreme pH can denature the enzyme molecules.

Skills

Investigating the effects of changes in temperature and pH on enzyme activity

You need to be able to describe how you could investigate the effects of changes in temperature and pH on enzyme activity. Part of this description needs to include how you would plan the investigation. In your plan, you should:
- suggest the most appropriate apparatus or technique and justify the choice made
- identify the independent variable and dependent variable

- describe how and explain why variables should be controlled
- suggest an appropriate number and range of values for the independent variable
- describe experimental procedures
- identify risks and describe and explain safety precautions
- describe and show how to record and process the results of an experiment to form a conclusion or to evaluate a prediction
- make reasoned predictions of expected results

Revision activity

Make a table of the enzymes you have learned about. Use the headings shown below.

Name of enzyme	Substrate (what the enzyme works on)	End product(s)	Other details (e.g. where the reaction happens, optimum pH)

Enzymes are specific

Enzymes are very specific (each chemical reaction is controlled by a different enzyme, so the enzyme has a high **specificity**). This is because of the shape of the **active site**, which has a complementary shape to the substrate (they fit together). The shape of the active site of protease will be different from the shape of the active site of amylase, for example. This means that protease cannot break down starch and amylase cannot break down proteins. In other words, the enzyme is specific. For an enzyme to catalyse a reaction, the enzyme molecule and the **substrate** molecule need to meet and join together by means of a temporary bond. This temporary structure is called the **enzyme–substrate complex**. The **product** or products are then released and the enzyme molecule can combine with another substrate molecule.

Enzymes and temperature

Enzymes work very slowly at low temperatures. This is because they have a low kinetic energy, so there are few collisions with the substrate molecules.

As the temperature is increased, the reaction rate increases because kinetic energy and, therefore, the rate of effective collisions increases. However, above the optimum temperature for the enzyme, the reaction rate starts to decrease. This is because enzyme molecules start to permanently lose their shape at high temperature. This deforms the active site, so the enzyme and substrate cannot fit together (so no reaction). This effect is called denaturing. Most enzymes are **denatured** above 50°C.

Enzymes are not denatured by low temperatures – they are just slowed down, and will work again when the temperature is suitable. Once an enzyme is denatured, the damage is permanent.

Enzymes and pH

An enzyme works best at its optimum pH. At a higher or lower pH, the enzyme is less effective and an extreme pH can denature the enzyme – the active site is deformed permanently. This means there is no longer a complementary fit between the enzyme molecule and the substrate.

Skills

Plotting graphs
Enzyme questions often involve plotting a set of results on a graph. Remember the key points for drawing a graph:
- Plot the independent variable (the figures you control) on the *x*- (horizontal) axis – these are the figures that usually go up in even stages.
- Label both axes with a title and units.
- Plot points in pencil (that way, you can change any mistakes).
- Join the points with a line (this can be a curve).

If you are instructed to predict a result using a graph, draw on the graph to read off the answer. Remember to state the units.

Sample question

Figure 5.2 shows a box of biological washing powder. Study the information on the box.

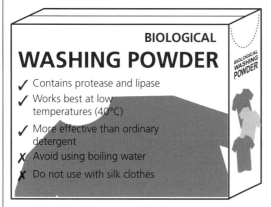

BIOLOGICAL WASHING POWDER
- ✓ Contains protease and lipase
- ✓ Works best at low temperatures (40°C)
- ✓ More effective than ordinary detergent
- ✗ Avoid using boiling water
- ✗ Do not use with silk clothes

▲ **Figure 5.2**

a Explain why:
  i the presence of protease and lipase would make the washing powder more effective than ordinary detergent [3]
  ii the powder should not be used in boiling water [2]
b Silk is a material made from protein. Explain why biological washing powder should not be used to wash silk clothes. [2]

Student's answer

> a i Protease and lipase are enzymes ✓, so they would break down stains ✓ better than ordinary detergent.
>   ii You could burn your hands when taking the clothes out. ✗
> b There is protease ✓ in the biological washing powder. This would digest the protein ✓ in the silk, so the clothes would get spoiled.

Teacher's comments

There are 3 marks available in part a(i). The student has made two valid statements, but has not given enough detail about what the enzymes digest (protease breaks down protein; lipase breaks down fats). In part a(ii) the student does not answer the question; the statement needs to be related to the properties of enzymes – they are denatured at high temperatures. The students produces a good response to part b, gaining both the available marks.

Correct answer

a i Protease and lipase are enzymes, so they break down stains better than ordinary detergent. Protease breaks down protein and lipase breaks down fat to form products that are soluble.
   ii Boiling water should not be used because it will denature the enzymes in the washing powder so they will not be able to break down stains.
b The protease in the biological washing powder would digest the protein in the silk, weakening the material, and thus spoiling the clothes.

Exam-style questions

1 Plan an investigation into the effect of pH on a sample of the enzyme amylase. [10]

2 Six identical samples containing a mixture of starch and amylase in water were kept at different temperatures, and the time taken for the starch to be digested was measured. The results are shown in the following table:

Temperature at which samples were kept/°C	Time for starch to be digested/ minutes
15	32
20	18
30	7
35	3
40	10
50	35

a Plot a graph of these results. [3]

b i   Describe how the time taken for the starch to be digested could be determined. [2]

   ii At what temperature was the starch digested most rapidly? [1]

   iii Describe the relationship between temperature and the rate of starch digestion. [2]

c Similar samples were set up and kept at 10°C and 60°C. The starch in these samples was not fully digested after 1 hour. Both of these samples were then kept at 35°C. Suggest and explain the effects of the following changes in temperature on starch digestion:

   i   Sample changed from 10°C to 35°C. [2]

   ii Sample changed from 60°C to 35°C. [2]

3 a Adult female mosquitoes feed on the blood of mammals. They produce a protein-digesting enzyme called trypsin.

   i   Explain why an adult female mosquito would need trypsin. [2]

   ii State the product that would be present in the gut of the mosquito if trypsin had been active. [1]

   iii State one use of this product in the body of an organism such as a mosquito. [1]

Scientists have found a way of introducing a hormone into mosquitoes to switch off the trypsin secretion.

   b Suggest how this treatment would affect adult female mosquitoes. [2]

   c Enzymes such as trypsin are easily damaged. Suggest two ways of damaging an enzyme. [2]

6 Plant nutrition

Key objectives

The objectives for this chapter are to revise:
- definitions of the key terms
- the word equation for photosynthesis
- the role of chlorophyll in chloroplasts
- the use and storage of carbohydrates made in photosynthesis
- the importance of nitrate and magnesium ions
- the need for chlorophyll, light and carbon dioxide for photosynthesis
- the effects of varying light intensity, carbon dioxide concentration and temperature on the rate of photosynthesis

- the effect of light and dark conditions on gas exchange in an aquatic plant
- how the surface area and the thin nature of leaves are adaptations for photosynthesis
- the cellular and tissue structure of a leaf
- how the internal structure of a leaf is adapted for photosynthesis

- the balanced chemical equation for photosynthesis
- the limiting factors of photosynthesis in different environmental conditions

Key terms

Term	Definition
Photosynthesis	The process by which plants synthesise carbohydrates from raw materials using energy from light

Photosynthesis

The word equation for **photosynthesis** is:

$$\text{carbon dioxide} + \text{water} \xrightarrow[\text{chlorophyll}]{\text{light}} \text{glucose} + \text{oxygen}$$

The equation shows that the raw materials for photosynthesis are carbon dioxide, water and light energy. The products are glucose (starch) and oxygen.

The glucose produced is converted to starch for storage in the leaf.

> For the extended paper, you need to learn the balanced chemical equation for photosynthesis:
>
> $$6CO_2 + 6H_2O \xrightarrow[\text{chlorophyll}]{\text{light}} C_6H_{12}O_6 + 6O_2$$

The process of photosynthesis

Green plants take in carbon dioxide through their leaves. This happens by diffusion.

Water is absorbed through the plants' roots by osmosis and transported to the leaf through xylem vessels.

Chloroplasts, containing a green pigment called chlorophyll, are responsible for transferring light energy into chemical energy for the synthesis of carbohydrates. This energy is used to break up water molecules and then to bond hydrogen and carbon dioxide to form glucose. This is usually changed to sucrose for transport around the plant.

The use of carbohydrates by the plant include:

- starch as an energy store; starch is insoluble and causes no osmotic problems
- sucrose for transport in the phloem
- cellulose to build cell walls; all plant cells have cellulose cell walls
- glucose, used in respiration to provide energy; this is needed for processes such as active uptake of mineral ions by the roots
- nectar in the flowers to attract insects for pollination

Oxygen is released as a waste product, or used by the plant for respiration.

Sample question [REVISED]

The chemical equation for photosynthesis shown below is incomplete:

$$6H_2O + \underline{\quad\quad} \xrightarrow[\text{plant pigment}]{\text{energy}} C_6H_{12}O_6 + \underline{\quad\quad}$$

a Complete the equation in either all words or all symbols. [2]
b State the source of energy for this reaction. [1]
c Name the plant pigment necessary for this reaction. [1]

Student's answer

a

$$6H_2O + 6CO_2 ✓ \xrightarrow[\text{plant pigment}]{\text{energy}} C_6H_{12}O_6 + \text{oxygen} ✗$$

b The Sun ✗
c Chloroplast ✗

Teacher's comments

Although the student has identified both compounds missing from the equation in part a, they have written one in symbols and the second in words, so the teacher has not awarded the second mark. When you write an equation or complete one, always do it with either words *or* symbols – do not mix them up. The correct answer here is $6O_2$. 'The Sun' is not specific enough for part b – the Sun produces two types of energy (light and heat). Plants use only light energy in photosynthesis. The correct answer was light or sunlight. The correct answer to part c is chloro*phyll*. Chloroplasts are structures, not pigments. They contain the pigment.

Factors needed for photosynthesis

You need to be able to describe how you would investigate the need for chlorophyll, light and carbon dioxide for photosynthesis, and identify the controls needed.

In these investigations, the plant used is tested for starch, since starch is the end-product of the process.

Skills

Testing a leaf for starch

Starch is stored in plant leaves as a product of photosynthesis. The starch test does not work by placing iodine solution on fresh leaves – it is not absorbed.

You need to be able to describe the starch test and the reasons for each stage. There are also some important safety points, outlined in Table 6.1.

▼ **Table 6.1 Testing a leaf for starch – stages and safety points**

Stage	Reason	Safety points
Boil the leaf in water	To kill the leaf – this makes it permeable	Danger of scalding
Boil the leaf in ethanol	To decolourise the leaf – chlorophyll dissolves in ethanol	No naked flames – ethanol is highly flammable
Rinse the leaf in water	Boiling the leaf in ethanol makes it brittle – the water softens it	
Spread the leaf out on a white tile	So that the results are easy to see	
Add iodine solution to the leaf	To test for the presence of starch	Avoid skin contact with iodine solution

At the start of the investigations, the plant is destarched. This involves leaving the plant in the dark for 48 hours. The plant uses up all the stores of starch in its leaves. One plant (or leaf) is exposed to all the conditions needed – this is the **control**. Another plant (or leaf) is deprived of one condition (this may be light or carbon dioxide).

Investigating the relationship with chlorophyll

Some plants have variegated leaves – only some parts of each leaf contain chlorophyll. When tested for starch, only the parts of the leaf with chlorophyll will contain starch.

Investigating the relationship with carbon dioxide

Figure 6.1 shows how the plant is set up to investigate if carbon dioxide is needed for photosynthesis. The carbon dioxide around a plant can be controlled by keeping the plant in a sealed container with a carbon dioxide absorber, such as sodium hydroxide pellets or sodium hydrogencarbonate solution. The control plant would be in an identical container, without the carbon dioxide absorber. After a few hours, the starch test is carried out on the control and test plant/leaf.

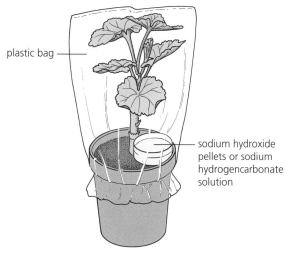

plastic bag

sodium hydroxide pellets or sodium hydrogencarbonate solution

▲ **Figure 6.1 Experimental set-up to investigate if carbon dioxide is needed for photosynthesis**

Investigating the relationship with oxygen production

It is difficult to show that land plants produce oxygen during photosynthesis because the gas diffuses into the air. However, some aquatic plants produce bubbles of oxygen. These can be collected and tested with a glowing splint – this relights in oxygen (Figure 6.2).

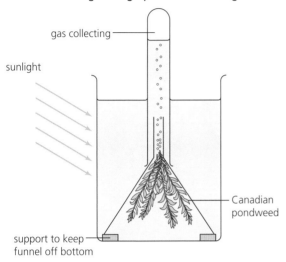

▲ **Figure 6.2 Experimental set-up to investigate if oxygen is produced during photosynthesis**

Investigating the relationship with light

The relationship between light intensity and the rate of photosynthesis can be demonstrated as shown using an aquatic plant such as *Elodea* (Canadian pondweed) – as the rate of photosynthesis increases, the rate of bubble production also increases.

The light intensity is related to the distance (*D*) between the lamp and the plant:

$$\text{light intensity} = \frac{1}{D^2}$$

As the lamp is moved closer, the light intensity increases. The rate of photosynthesis is directly proportional to the light intensity, as shown by the graph in Figure 6.3.

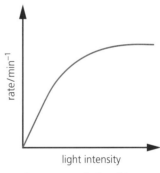

▲ **Figure 6.3 Relationship between light intensity and the rate of photosynthesis**

Skills

Investigating the effect of light and dark on gaseous exchange in an aquatic plant

Three tubes are set up as shown in Figure 6.4 and Table 6.2.

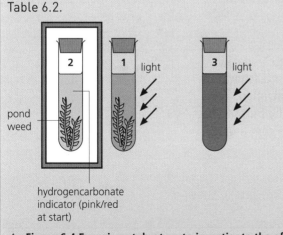

▲ **Figure 6.4 Experimental set up to investigate the effect of light and dark on gaseous exchange**

▼ **Table 6.2 Experimental conditions for each tube**

Tube	Contents	Treatment
1	Hydrogencarbonate indicator solution + aquatic plant	Exposed to light
2	Hydrogencarbonate indicator solution + aquatic plant	Kept in the dark
3 (control)	Hydrogencarbonate indicator solution only	Exposed to light

Hydrogencarbonate indicator solution is used to test for the presence of carbon dioxide. An increase in CO_2 levels turns it from pink/red to yellow. A decrease in CO_2 levels turns it purple.

The results and conclusions are shown in Table 6.3.

▼ **Table 6.3 Expected results and conclusions for each tube**

Tube	Colour of hydrogencarbonate indicator solution	Conclusion
1	Purple	The plant has taken in CO_2 for photosynthesis
2	Yellow	The plant cannot photosynthesise in the dark, but has continued to respire, producing CO_2
3 (control)	Pink	There was no plant present, so there was no change in levels of CO_2

Photosynthesis only happens when there is light, but respiration happens all the time (in both plants and animals).

Limiting factors

When light intensity, carbon dioxide concentration or temperature is increased, the rate of photosynthesis increases. The photosynthetic rate cannot be increased indefinitely, however.

A **limiting factor** is something present in the environment in such short supply that it limits life processes.

Looking at Figure 6.3, there comes a point when further increases in light intensity do not increase the rate – a point is reached where all the chloroplasts cannot trap any more light. If there is another limiting factor, such as carbon dioxide concentration, the rate of photosynthesis will also be limited, as shown in Figure 6.5.

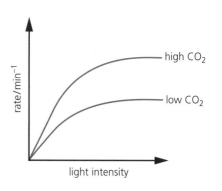

▲ **Figure 6.5 Effect of carbon dioxide levels on the rate of photosynthesis**

Environmental conditions and photosynthesis

Environmental conditions vary in many parts of the world according to the season. In the winter months, plants may experience much lower temperatures and lower light intensity, with shorter days, while the opposite is true in the summer months. Some plants have become adapted to shade conditions, or complete their life cycle in spring before larger plants develop leaves and create shade.

Glasshouses and polytunnels can be used control conditions for plant growth, especially when growing conditions outside are not ideal. The glass helps trap heat inside and atmospheric conditions can be controlled.

Carbon dioxide enrichment

Atmospheric air contains only 0.04% carbon dioxide, so it can easily become a factor that limits the rate of photosynthesis. A glasshouse is a closed system, so the content of the air in it can be controlled. For example, the amount of carbon dioxide can be increased by burning fossil fuels in the greenhouse, or releasing pure carbon dioxide from a gas cylinder.

Optimum light

If light conditions in a glasshouse are not optimum (e.g. in winter), they can be improved by using artificial lights.

Optimum temperature
If the temperature is a limiting factor, for example in winter, it can be raised by using a heating system. If fossil fuels are burned, there is also a benefit from the carbon dioxide produced.

Mineral requirements

Nitrate ions are needed for making amino acids. These are the building blocks of proteins. Remember that all proteins contain the element nitrogen (see Chapter 4). Each amino acid is formed by combining sugars, made during photosynthesis, with nitrate. The amino acids are made into long chains by bonding them together. The proteins are used to make cytoplasm and enzymes.

Magnesium ions are needed to make chlorophyll. Each chlorophyll molecule contains one magnesium atom. Plants need chlorophyll to trap light to provide energy during photosynthesis.

Leaf structure

You need to be able to identify the cellular and tissue structure of a leaf of a dicotyledonous plant (Figure 6.6).

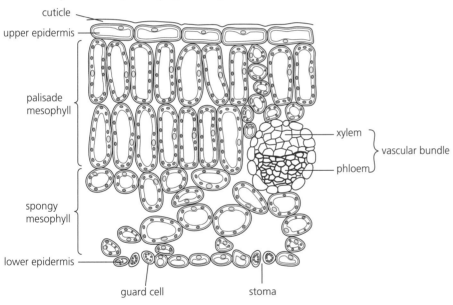

▲ **Figure 6.6 A cross-section through part of a leaf**

Adaptation of leaves for photosynthesis

- Their broad, flat shape offers a large surface area for absorption of sunlight and carbon dioxide.
- Most leaves are thin, so the carbon dioxide only has to diffuse across short distances to reach the inner cells.
- The large spaces between cells inside the leaf provide an easy passage through which carbon dioxide can diffuse.
- There are many stomata (pores) in the lower surface of the leaf. These allow the exchange of carbon dioxide and oxygen with the air outside.

Revision activity

Practise drawing, labelling and annotating diagrams:
- Copy the diagram of a leaf in Figure 6.6, for example by tracing.
- Practise labelling the structures present.
- Annotate the diagram by writing one statement about each of the labels.

- There are more chloroplasts in the upper (palisade) cells than in the lower (spongy mesophyll) cells. The palisade cells, being on the upper surface, will receive most sunlight and this will reach the chloroplasts without being absorbed by too many cell walls.
- The branching network of veins provides a good water supply to the photosynthesising cells.

Parts of the leaf and their functions

Table 6.4 explains how each feature of the internal structure of a leaf is adapted for photosynthesis.

▼ Table 6.4 Adaptations of the internal structure of a leaf for photosynthesis

Part of leaf	Details
Cuticle	Made of wax, it waterproofs the leaf
	It is secreted by cells of the upper epidermis
Upper epidermis	Thin and transparent cells that allow light to pass through
	No chloroplasts are present
	Acts as a barrier to disease organisms
Palisade mesophyll	Main region for photosynthesis
	Cells are columnar (quite long) and packed with chloroplasts to trap light energy
	They receive carbon dioxide by diffusion from air spaces in the spongy mesophyll
Spongy mesophyll	Cells are more spherical and loosely packed
	They contain chloroplasts, but not as many as in palisade cells
	Air spaces between cells allow gaseous exchange – carbon dioxide to the cells, oxygen from the cells – during photosynthesis
Vascular bundle	A leaf vein made up of xylem and phloem
	Xylem vessels bring water and mineral ions to the leaf
	Phloem vessels transport sugars and amino acids away (this is called translocation)
Lower epidermis	Acts as a protective layer
	Stomata are present to regulate the loss of water vapour (this is called transpiration)
	Site of gaseous exchange into and out of the leaf
Stomata	Each stoma is surrounded by a pair of guard cells
	These can control whether the stoma is open or closed
	Water vapour passes out during transpiration
	Carbon dioxide diffuses in and oxygen diffuses out during photosynthesis

Sample question

Match the parts of a leaf to their functions. Functions may be linked once or more than once. [5]

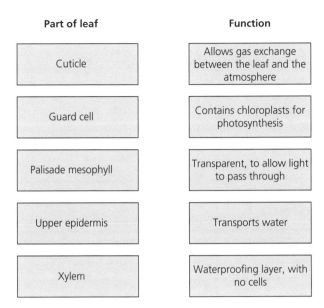

Part of leaf	Function
Cuticle	Allows gas exchange between the leaf and the atmosphere
Guard cell	Contains chloroplasts for photosynthesis
Palisade mesophyll	Transparent, to allow light to pass through
Upper epidermis	Transports water
Xylem	Waterproofing layer, with no cells

Student's answer

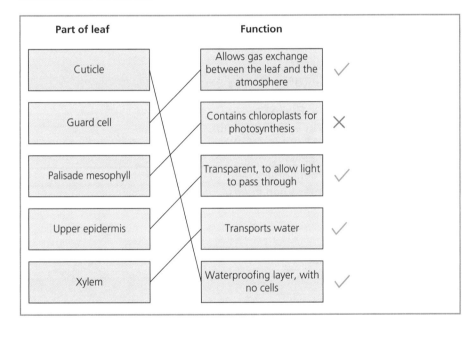

Part of leaf	Function	
Cuticle	Allows gas exchange between the leaf and the atmosphere	✓
Guard cell	Contains chloroplasts for photosynthesis	✗
Palisade mesophyll	Transparent, to allow light to pass through	✓
Upper epidermis	Transports water	✓
Xylem	Waterproofing layer, with no cells	✓

Teacher's comments

This gains four of the five marks available. The reason the mark was not awarded for the 'contains chloroplasts for photosynthesis' function was because palisade mesophyll *and* guard cells both have this function – both were needed for the mark.

Exam-style questions

1 Figure 6.7 shows a variegated leaf in a photosynthesis experiment. Part of the leaf had been covered with black paper. The leaf was then exposed to light for a few hours. Leaf discs were then cut from regions of the leaf at A, B, C and D. Each disc was tested for the presence of starch. Predict the results of the starch test on each region of the variegated leaf. Give a reason for each result. [8]

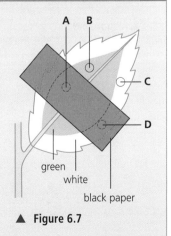

green
white
black paper

▲ Figure 6.7

2 Figure 6.8 shows the rates of sugar production by a plant on a bright day and on a dull day.

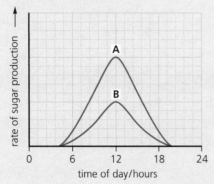

▲ **Figure 6.8**

a i Which curve, A or B, shows sugar production on a bright day? State a reason for your choice. [2]

ii Outline the role of chloroplasts in photosynthesis. [2]

iii Suggest one feature of a leaf, other than the presence of chloroplasts, that might affect the amount of sugar produced. [1]

b i Suggest why many plants store starch rather than sugars. [1]

ii Which chemical reagent is used to test for starch? [1]

3 A student carried out an experiment to investigate the growth of floating water plants taken from a pond. Equal masses of the plants were placed into three separate glass containers – A, B and C – similar to that shown in Figure 6.9. Container A was lit by a 250 W bulb, container B was lit by a 75 W bulb and container C was lit by a 250 W bulb with a coloured filter in front of the lamp. At weekly intervals, the plants were removed from each container in turn, gently dried, weighed and returned to their container after their mass had been recorded. Figure 6.10 shows the results.

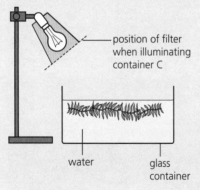

▲ **Figure 6.9**

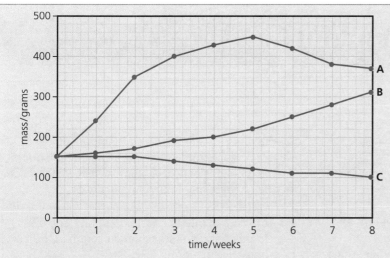

▲ Figure 6.10

a Calculate the percentage increase in mass of the plants in container A during the first 5 weeks of the experiment. (Show your working.) [2]

b Suggest why the mass of plants in container A began to decrease after week 5, while the mass of plants in container B continued to increase. [2]

c During the eighth week, in which container would there be the least dissolved oxygen? Explain your answer. [2]

Figure 6.11 shows the amount of light of different colours absorbed by chlorophyll. The filter used in illuminating container C allowed only one colour of light to pass through to the water plants.

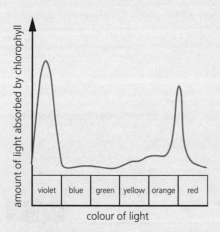

▲ Figure 6.11

d Suggest which colour of light passed through the filter. Explain your answer. [2]

4 a i Name four tissues found in a leaf. [4]
   ii For each named tissue, state its function. [4]

b Explain why a plant with a magnesium deficiency would not grow as well as a plant with an unlimited magnesium supply. [2]

7 Human nutrition

Key objectives

The objectives for this chapter are to revise:
- definitions of the key terms
- the main dietary sources and the importance of the main foodstuffs
- the causes of scurvy and rickets
- the main organs of the digestive system and their functions
- the processes involved in physical digestion
- the types and structure of human teeth and their functions
- the significance of chemical digestion in the digestive system
- the sites of secretion and functions of key enzymes, listing substrates and end products
- the functions of hydrochloric acid in gastric juice
- the absorption of nutrients and water

- the role of bile in emulsifying fats and oils
- the digestion of starch and protein in the digestive system
- the significance of villi in the small intestine, their structure and the role of capillaries and lacteals

Key terms

Term	Definition
Absorption	The movement of nutrients from the intestines into the blood
Assimilation	The uptake and use of nutrients by cells
Balanced diet	Diet that contains all the essential nutrients in the correct proportions to maintain good health
	The nutrients needed are carbohydrate, fat, protein, vitamins, minerals ions, fibre (roughage) and water
Chemical digestion	The breakdown of large insoluble molecules into small soluble molecules
Digestion	The breakdown of food
Egestion	The passing out of food that has not been digested or absorbed, as faeces through the anus
Ingestion	The taking of substances (e.g. food, drink) into the body through the mouth
Physical digestion	The breakdown of food into smaller pieces without chemical change to the food molecules

Diet

A **balanced diet** must contain enough carbohydrates and fats to meet our energy needs. It must also contain enough protein to provide the essential amino acids to make new cells and tissues for growth or repair. The diet must also include vitamins and mineral ions, plant fibre (roughage) and water (Table 7.1).

Vitamins and minerals, although needed in only small quantities, are important for maintaining good health. A shortage can result in a deficiency disease. You only need to know vitamins C and D, and the minerals calcium and iron. Fibre (roughage) is needed in much larger quantities. Do not forget that water is also a vital part of our dietary requirements.

Revision activity

Make a mnemonic to remember all of the nutrients required in a balanced diet. You need to use the letters C, F, P, V, M, F, W.

▼ **Table 7.1 Classes of food**

Nutrient	Importance in the body	Good food sources
Carbohydrate	Source of energy	Rice, potato, sweet potato, cassava, bread, millet, sugary foods (e.g. cake, jam, honey)
Fat/oil (fats are solid at room temperature, oils are liquid)	Source of energy (twice as much as carbohydrate); used as insulation against heat loss, for some hormones, in cell membranes, for insulation of nerve fibres	Butter, milk, cheese, egg yolk, animal fat, groundnuts (peanuts)
Protein	Growth, tissue repair, enzymes, some hormones, cell membranes, hair, nails; can be broken down to provide energy	Meat, fish, eggs, soya, groundnuts, milk, meat substitute (e.g. Quorn), cowpeas, falafel
Vitamin C	Needed to maintain healthy skin and gums; a deficiency can lead to scurvy	Citrus fruits, blackcurrants, cabbage, tomato, guava, mango
Vitamin D	Needed to maintain hard bones; helps in **absorption** of calcium from small intestine; deficiency can lead to rickets	Milk, cheese, egg yolk, fish liver oil; can be made in the skin when exposed to sunlight
Iron	Needed for formation of haemoglobin in red blood cells; a deficiency can lead to anaemia	Red meat, liver, kidney, eggs, green vegetables (spinach, cabbage, cocoyam, groundnut leaves), chocolate
Calcium	Needed to form healthy bones and teeth, and for normal blood clotting; a deficiency can lead to rickets	Milk, cheese, fish
Dietary fibre (roughage)	This is cellulose, which adds bulk to undigested food passing through the intestines, maintaining peristalsis; a deficiency can lead to constipation	Vegetables, fruit, wholemeal bread
Water	Formation of blood, cytoplasm, as a solvent for transport of nutrients and removal of wastes (as urine); enzymes work only in solution	Drinks, fruit, vegetables

The causes of scurvy and rickets

A shortage of **vitamin C** can lead to a deficiency disease called scurvy. Fibres in connective tissue of skin and blood vessels do not form properly, leading to bleeding under the skin. Other symptoms are feeling constantly tired, weak and irritable, with joint pains and swollen, bleeding gums. In severe cases, the teeth can fall out.

A shortage of **vitamin D** or **calcium** can lead to a deficiency disease called rickets. The symptoms are soft bones that become deformed. Sufferers may become bow legged. Exposure to moderate sunlight helps the body make vitamin D. Thus, a lack of exposure (because of climate or season, or wearing clothing that acts as a barrier to sunlight) can result in the development of a deficiency.

Sample question

REVISED

The following table shows the carbohydrate content of some vegetables.

Vegetable	Total carbohydrate/g per 100 g	Starch/g per 100 g	Fibre (roughage)/g per 100 g
Beans	15.1	9.3	3.5
Broccoli	1.1	Trace	2.3
Cabbage	4.1	0.1	2.4
Carrots (boiled)	4.9	0.2	2.5

Vegetable	Total carbohydrate/g per 100 g	Starch/g per 100 g	Fibre (roughage)/g per 100 g
Chickpeas	18.2	16.6	4.3
Onions	3.7	Trace	0.7
Peas (frozen, boiled)	9.7	4.7	5.1
Potato (boiled)	17.0	16.3	1.2
Sweet potato (boiled)	20.5	8.9	2.3
Tomatoes (raw)	3.1	Trace	1.0

a Name the chemical elements present in a carbohydrate. [1]
b State which vegetable in the table contains:
  i   the highest proportion of total carbohydrate [1]
  ii the highest proportion of fibre (roughage). [1]
c Total carbohydrate is calculated as the sum of starch and sugars in the vegetable.
  i   Name the vegetable that contains the highest proportion of sugar per 100 g vegetable. [1]
  ii Calculate the amount of sugar present in 500 g of the vegetable named in c(i). Show your working. [2]

Student's answer

a   C, H, O ✗
b i   Sweet potato ✓
   ii Peas ✓
c i   Sweet potato ✓
   ii 20.5 – 8.9 = 11.6
       11.6 × 5 ✓ = 58 ✗

Teacher's comments

Most of the answers were good, but this student made two easily avoidable errors:
  a   Do not use abbreviations such as symbols when you are asked to name elements.
c(ii) Remember to state the units when giving the answer to a calculation. This answer gained 1 mark for showing the correct working for the calculation, but lost the second mark because of the lack of units – which should have been 'g' (grams).

Skills

Manipulating data
You need to be able to apply your mathematical skills to manipulate data in some Biology exam questions. For example, in the sample question above there is data about the carbohydrate content of some vegetables. The values are in g/100 g. For data like this, you may be asked to calculate the total amount of one of the groups of carbohydrates in a given amount of vegetable.

If the question asked 'How much fibre is present in 30 g of broccoli?':

- First read off the amount of fibre in 100 g. *This is 2.3 g.*
- Divide 2.3 g by 100 (to find the amount of fibre in 1 g of broccoli):

$$\frac{2.3}{100} = 0.023\,g$$

- Then multiply this value by 30:

$$0.023 \times 30 = 0.69$$

- Remember to include the units in your answer. *There are 0.69 g of fibre in 30 g of broccoli.*

Digestive system

Figure 7.1 shows the main organs of the digestive system. Table 7.2 gives their functions.

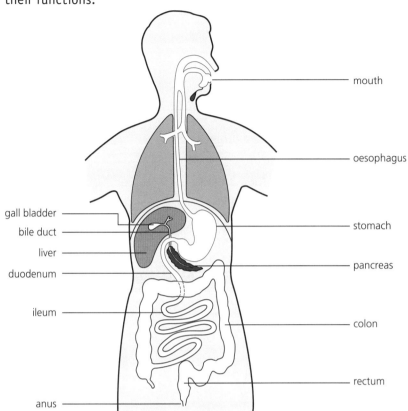

▲ **Figure 7.1 The human digestive system**

Revision activity

Make a list of all the main parts of the digestive system. For each part, write down one function.

▼ **Table 7.2 Regions of the digestive system and their functions**

Region	Function
Mouth	Food is **ingested** here. It is mechanically digested by cutting, chewing and grinding of teeth. Saliva is added.
Salivary glands	Produce saliva containing the enzyme amylase to begin the **chemical digestion** of starch. The water in saliva helps lubricate food and makes small pieces stick together.
Oesophagus	Boluses (balls) of food pass through, from mouth to stomach.
Stomach	Muscular walls squeeze the food to make it semi-liquid. Gastric juice contains protease to chemically digest protein and hydrochloric acid to maintain an optimum pH (1–2.5). The acid also kills bacteria.
Duodenum	The first part of the small intestine. It receives pancreatic juice containing protease, lipase and amylase. The juice also contains sodium hydrogencarbonate, which neutralises stomach acid, giving a pH of 7–8. Carbohydrates, fats and proteins are digested here.
Ileum	The second part of the small intestine. Enzymes in the epithelial lining chemically digest maltose and peptides. Its surface area is increased by the presence of villi, which allow the efficient absorption of digested food molecules. Most water is reabsorbed here.
Pancreas	Secretes pancreatic juice into the duodenum for chemical digestion of proteins, fats and starch.
Liver	Makes bile, which is stored in the gall bladder. Note that the liver does not make digestive enzymes – bile is not an enzyme. Bile contains salts that emulsify fats, forming droplets with a large surface area to make **digestion** by lipase more efficient. It does not change the fat molecules chemically – it is just the droplet size that changes from large to small due to the action of bile. Digested foods are **assimilated** here – for example, glucose is stored as glycogen. Surplus amino acids are deaminated (see Chapter 13).
Gall bladder	Stores bile, made in the liver, to be secreted into the duodenum via the bile duct.

Region	Function
Colon	The second part of the large intestine. It reabsorbs water from undigested food, resulting in the formation of faeces. It also absorbs bile salts to pass back to the liver.
Rectum	Stores faeces until they are **egested**.
Anus	Has muscles to control when faeces are egested from the body.

Physical digestion

Physical digestion increases the surface area of food for the action of enzymes in chemical digestion. It does not involve breaking down large molecules into small molecules.

Teeth are involved in the physical digestion of food. Chewing food breaks down large pieces into smaller pieces, giving a larger surface area for enzymes to work on. You need to know about the types and functions of human teeth, and also about tooth structure. Figure 7.2 shows the types and functions of human teeth.

	Incisor	Canine	Premolar	Molar
Position in mouth	Front	Either side of incisors	Behind canines	Back
Description	Chisel-shaped (sharp edge)	Slightly more pointed than incisors	Have two points (cusps). Have one/two roots	Have four/five cusps. Have two/three roots
Function	Biting off pieces of food	Similar function to incisors	Tearing and grinding food	Chewing and grinding food

▲ **Figure 7.2 Summary of types of human teeth and their functions**

Tooth structure

Figure 7.3 shows a section through a molar tooth.

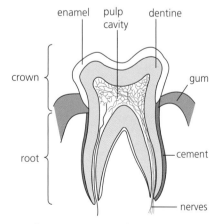

▲ **Figure 7.3 Section through a molar tooth**

Function of the stomach in physical digestion

The muscular walls of the stomach squeeze on food to make it semi-liquid. This increases the surface area of the food, and the churning action brings the food into contact with enzyme molecules.

Hydrochloric acid in gastric juice (in the stomach)

You need to be aware of the functions of hydrochloric acid in gastric juice.

- The hydrochloric acid that is secreted by cells in the wall of the stomach creates a very acid pH of 2. This provides the optimum pH for the protein-digesting enzyme, protease, to work.
- The low pH is also important because it denatures enzymes in harmful organisms in food, such as bacteria (which could cause food poisoning).

The role of bile

Bile is made in the liver, stored in the gall bladder and transferred to the duodenum by the bile duct (Figure 7.1). It has no enzymes but does contain bile salts, which act on fats in a similar way to a detergent.

The bile salts emulsify the fats, breaking them up into small droplets with a large surface area, which are more efficiently digested by lipase.

Bile is slightly alkaline, as it contains sodium hydrogencarbonate, and has the function of neutralising the acidic mixture of food and gastric juices as it enters the duodenum. This is important because the enzymes secreted into the duodenum need alkaline conditions to work at their optimum rate.

Chemical digestion

REVISED

Food that we ingest is mainly made up of large, insoluble molecules that cannot be absorbed through the gut wall. It needs to be changed into small, soluble molecules through the process of chemical digestion.

Enzymes speed up the process. They work efficiently at body temperature (37°C) and at a suitable pH. The main places where chemical digestion happens are the mouth, stomach and small intestine. You need to be able to state the functions of amylase, protease and lipase (Table 7.3).

▼ Table 7.3 Amylase, protease and lipase

Enzyme	Site of action	Special conditions	Substrate digested	End product(s)
Amylase	Mouth, duodenum	Slightly alkaline	Starch	Simple reducing sugars, e.g. maltose, glucose
Protease	Stomach, duodenum	Acid in stomach, alkaline in duodenum	Protein	Amino acids
Lipase	Duodenum	Alkaline	Fats and oils	Fatty acids and glycerol

Digestion of starch

Starch is digested in two places in the digestive system: by salivary amylase in the mouth and by pancreatic amylase in the duodenum. **Amylase** works best in a neutral or slightly alkaline pH and converts large, insoluble starch molecules into smaller, soluble maltose molecules. Maltose is a disaccharide sugar and is still too big to be absorbed through the wall of the intestine. Maltose is broken down to glucose by the enzyme **maltase**, which is present on the membranes of the epithelial cells of the villi in the small intestine.

Digestion of protein

There are several proteases that break down proteins. One protease is **pepsin**, which is secreted in the stomach. Pepsin acts on proteins in the acidic conditions of the stomach, and breaks them down into soluble compounds called peptides. These are shorter chains of amino acids than proteins. Another protease is **trypsin**. This is secreted by the pancreas in an inactive form, which is changed to an active enzyme in the duodenum (part of the small intestine). It has a similar role to pepsin, breaking down proteins to peptides, but trypsin works efficiently in the alkaline conditions of the duodenum.

Absorption

REVISED

The **small intestine** is the region where the majority of nutrients are absorbed. It has a very rich blood supply. Digested food molecules are small enough to pass through the wall of the intestine into the bloodstream.

The small intestine and the colon are both involved in the absorption of water, but the small intestine absorbs the most.

Role of villi in absorption

You need to be able to relate the structure of the small intestine to its function of absorbing digested food and describe the significance of **villi** in increasing the internal surface area.

Villi are finger-like projections in the small intestine that increase the surface area for absorption. If a section of small intestine was turned inside out, its surface would be like a carpet. The surface area of a villus is further increased by the presence of microvilli.

Inside each villus are **blood capillaries** that absorb amino acids and glucose. There are also **lacteals**, which absorb fatty acids and glycerol.

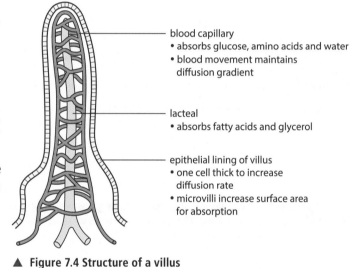

blood capillary
- absorbs glucose, amino acids and water
- blood movement maintains diffusion gradient

lacteal
- absorbs fatty acids and glycerol

epithelial lining of villus
- one cell thick to increase diffusion rate
- microvilli increase surface area for absorption

▲ **Figure 7.4 Structure of a villus**

Food molecules are absorbed mainly by diffusion. Figure 7.4 shows the features of a villus that increase the efficiency of diffusion. Molecules can also be absorbed by active transport (see Chapter 3). Epithelial cells contain mitochondria to provide energy for absorption against the concentration gradient.

Sample question

REVISED

a Proteins are digested in the stomach and small intestine.
  i Which type of enzyme breaks down proteins?   [1]
  ii State how the conditions necessary for the digestion of proteins in the stomach are different from those in the small intestine.   [1]

b When carbohydrates have been digested, excess glucose is stored.
  i Where is it stored? [1]
  ii What is it stored as? [1]
c Excess amino acids cannot be stored. Describe how they are
removed from the body. [4]

Student's answer

a	i	Protease ✓
	ii	It is acid in the stomach and alkaline in the small intestine. ✓
b	i	In the liver ✓
	ii	Glucagon ✗
c		The liver ✓ breaks them down. This makes urea. ✓ The kidney filters out the urea. ✓

Teacher's comments

The answers to parts a and b are good, except that the student has confused glucagon (a hormone) with glycogen (the correct answer). Names such as this have to be accurately spelt. The answer for part c contains only three statements. Further details about the filtering of blood or the formation of urine and its removal by urination would have achieved the final mark.

Exam-style questions

1 The chart in Figure 7.5 is used to find a person's ideal mass.

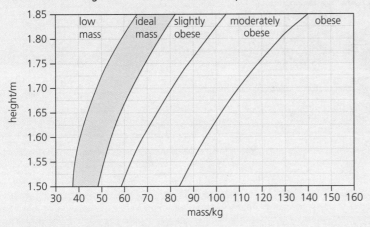

▲ **Figure 7.5**

The data in the following table were collected for three students, X, Y and Z.

Student	Mass/kg	Height/m
X	50.8	1.55
Y	63.8	1.85
Z	114.3	1.65

Identify the student who is:
  a obese   **b** of low mass   **c** of ideal mass [3]

2 Figure 7.6 shows the four types of teeth found in humans.

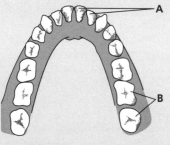

▲ **Figure 7.6**

a Copy the figure and label one example of each of the four types of teeth. [4]

b i What is the function of the teeth labelled A? [1]

ii What is the function of the teeth labelled B? [1]

c The outer layer of the crown of a tooth is resistant to attack by bacteria.

i Name this outer layer. [1]

ii State the mineral and the vitamin needed in the diet for the healthy development of this layer. [2]

3 Figure 7.7 shows an analysis of four food samples.

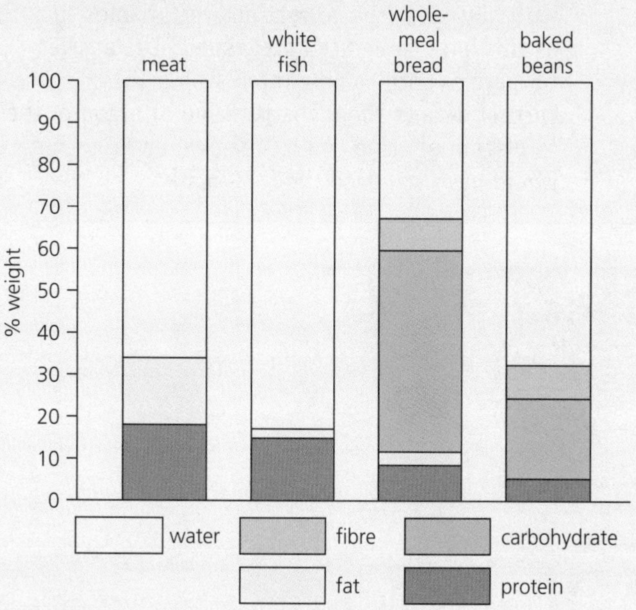

▲ Figure 7.7

a Which food contains:

i most protein [1]

ii most carbohydrate [1]

iii most water [1]

iv least fat? [1]

b (i) Suggest why meat contains no fibre. [2]

(ii) Explain why fibre is important in a balanced diet. [2]

c Where is water absorbed in the digestive system? [2]

d The graph does *not* contain data about all the components of a balanced diet. Which components are missing? [2]

4 Name the part(s) of the digestive system responsible for:

a moving food from the mouth to the stomach [1]

b producing bile [1]

c storing faeces [1]

d digesting starch [2]

e absorbing water [2]

f physical digestion [2]

g secreting three different enzymes [1]

Key objectives

The objectives for this chapter are to revise:
- definitions of the key terms
- the functions of xylem and phloem, and their positions in sections of roots, stems and leaves
- how to identify root hair cells and state their functions
- the pathway taken by water through root, stem and leaf through the xylem vessels
- the process of water loss from a leaf
- the effects of varying temperature and wind speed on transpiration rate

- how the structure of xylem vessels is related to their function
- how water vapour loss is related to the large surface area of cell surfaces, interconnecting air spaces and stomata
- the mechanisms of water movement through a plant and how and why wilting occurs
- how to explain the effects of variation of temperature, wind speed and humidity on transpiration rate
- the concepts of source and sink

Key terms

REVISED

Term	Definition
Transpiration	The loss of water vapour from leaves
Translocation	The movement of sucrose and amino acids in the phloem from sources to sinks

Xylem and phloem

REVISED

Two types of tissue are present in plants to transport materials. The **xylem** carries water and mineral ions, as well as providing support for the plant. The **phloem** carries food substances – sugars and amino acids.

You need to be able to identify xylem and phloem in sections of root (Figure 8.1), stem (Figure 8.2) and leaves (Figure 6.6, p. 39).

- Root – xylem is in the centre, forming a star, with phloem around it.
- Stem – the tissues are in oval-shaped vascular bundles, with xylem on the inside and phloem towards the outside.
- Leaves – the tissues are in vascular bundles, with xylem above the phloem.

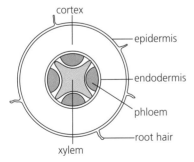

▲ **Figure 8.1 Section through a root**

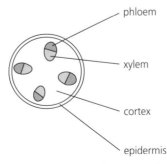

▲ **Figure 8.2 Section through a stem**

Structure and function of xylem vessels

Xylem cells (Figure 8.3) are long and thin, arranged end-to-end to form vessels (tubes). When mature, the cells lack cross walls, making a long, continuous tube. They lack cell contents such as cytoplasm and nuclei. The cell walls become lignified.

Their function is to transport water and mineral ions from roots to leaves. Lignin provides strength for the stem and makes the vessels waterproof.

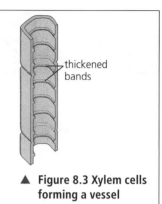

thickened bands

▲ **Figure 8.3 Xylem cells forming a vessel**

Water uptake

Root hair cells are tiny, tube-like outgrowths that form on young roots. They are very numerous and increase the surface area of the root for absorption of water and mineral ions. Figure 8.4 shows a root hair cell.

Skills

Investigating the pathway of water through a plant

You need to be familiar with how you would investigate the pathway of water through the above-ground parts of a plant. To do this you can use a plant shoot that has soft tissue (*Impatiens*, daffodil stalk with flower, or a stick of celery works well). The shoot is placed in a beaker of dilute stain, such as methylene blue or food colouring, and left for 24 hours. After the shoot has been removed from the dye, the stem can be sectioned with a sharp scalpel to observe which tissues have been stained by the dye (Figure 8.5). You may also

be able to see the veins in a daffodil flower that have picked up the blue dye.

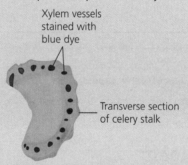

Xylem vessels stained with blue dye

Transverse section of celery stalk

▲ **Figure 8.5 A stem section stained with blue dye**

Passage of water through root, stem and leaf

Water passes through the root hair cells to the root cortex cells by osmosis, reaching the xylem vessels in the centre (Figure 8.1).

When water reaches the xylem, it travels up these vessels through the stem to the leaves. Mature xylem cells have no cell contents, so they act like open-ended tubes, allowing the free movement of water through them. In the leaves, water passes out of the xylem vessels into the surrounding cells (mesophyll cells). Mineral ions are also transported through the xylem.

▲ **Figure 8.4 A root hair cell**

Revision activity

Copy or trace the root hair cell in Figure 8.4. Label all its parts, remembering to use a ruler for all label lines.

Transpiration

Transpiration is the loss of water vapour from leaves. Water forms a thin layer on the surfaces of the leaf mesophyll cells and evaporates into the air spaces in the spongy mesophyll. This creates a high concentration of water molecules. If there is a higher concentration of water vapour inside the leaf than outside, water will diffuse out of the leaf through the stomata. Water evaporating from the leaves causes suction, which pulls water up the stem.

The loss of water vapour in transpiration is related to the large surface area provided by the cell surfaces in the mesophyll of the leaf, the interconnecting air spaces in the spongy mesophyll and the stomata.

Water movement in the xylem

Water vapour evaporating from a leaf creates a kind of suction (**transpiration pull**) because water molecules are held together by forces of attraction between them (**cohesion**). Therefore, the water forms a **column** and is drawn into the leaf from the xylem. This creates a transpiration stream, pulling water up from the root. Xylem vessels act like tiny tubes, drawing water up the stem by capillary action. Roots also produce a root pressure, forcing water up xylem vessels.

Water does *not* travel through xylem vessels by osmosis. Osmosis involves the movement of water across cell membranes – xylem cells do not have living contents when mature, so there will be no membranes.

Wilting

Young plant stems and leaves rely on their cells being turgid to keep them rigid. If the amount of water lost from the leaves of a plant is greater than the amount taken into the roots, the plant will have a water shortage. Cells become flaccid if they lack water, and they will no longer press against each other. Stems and leaves then lose their rigidity and wilt.

Sample question

REVISED

a Describe how the structure of xylem tissue is adapted to its functions. [3]
b Describe the mechanism of water movement through the xylem. [2]

Student's answer

a The cells join together to make a long ✓ tubular structure. There are no cross-walls ✓ and no living contents, ✓ so the water and mineral salts (✓) can pass through freely.

b Water moves by the pull from the leaves ✓ caused by transpiration ✓. Xylem vessels are very thin, so they act like a capillary tube (✓), helping to draw water upwards.

Teacher's comments

Both answers are excellent, gaining the maximum marks available. This student has learned the details of water transport in plants really well. The ticks in brackets mean that the statements are correct, but the maximum marks for the question have already been reached.

Rate of transpiration

Table 8.1 shows two factors that can affect the rate of transpiration. If these factors are reversed, the rate will also reverse.

▼ **Table 8.1 Factors that affect the rate of transpiration**

Factor	Effect on transpiration rate
Increase in temperature	Increases transpiration rate
Increase in wind speed	Increases transpiration rate

Most of the factors that result in a change in transpiration rate are linked to diffusion. When writing explanations, try to include references to the concentration gradient caused by a change in the factor.

Skills

Investigating the effect of temperature and wind speed on the rate of transpiration

You need to be familiar with how you would investigate the effects of variation of temperature and wind speed on transpiration rate. The apparatus used for transpiration investigations is a potometer. There are two types – the weight potometer (Figure 8.6) and the bubble potometer (Figure 8.7).

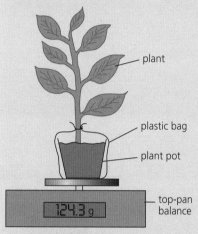

▲ **Figure 8.6 Weight potometer**

The weight potometer involves measuring the weight loss of a plant over time by setting up the plant on a top-pan balance. The plant pot is wrapped and sealed in a plastic bag so that the only water loss is from the plant. The plant can

be exposed to different conditions, for example changes in temperature or wind speed.

The bubble potometer involves recording the rate of uptake of water by a leafy shoot. The shoot is attached to a water-filled capillary tube, into which a bubble is introduced. The movement of the bubble is recorded over time. The shoot can be exposed to different conditions, for example changes in temperature or wind speed.

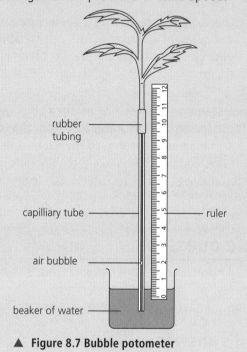

▲ **Figure 8.7 Bubble potometer**

- An increase in temperature increases the kinetic (movement) energy of the water molecules, so they diffuse faster. Transpiration is likely to be faster on a hot day than on a cold day.
- An increase in humidity lowers the transpiration rate. This is because it increases the concentration of water molecules outside the leaf, reducing the concentration gradient for diffusion.
- An increase in wind speed lowers the concentration of water molecules outside the leaf, increasing the concentration gradient for diffusion, so the transpiration rate increases.

Skills

Controlling variables in transpiration investigations

For the Practical and Alternative to Practical examination papers, you need to be able to describe how and why variables should be controlled. This can be demonstrated by the following transpiration investigation into which surface of a leaf loses more water. The apparatus is shown in Figure 8.8.

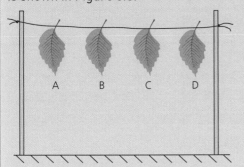

A – Both sides covered with petroleum jelly
B – Upper epidermis covered with petroleum jelly
C – Lower epidermis covered with petroleum jelly
D – Neither surface covered with petroleum jelly

▲ **Figure 8.8 Apparatus to investigate transpiration from different leaf surfaces**

Each of the four leaves has a different treatment, and there are a number of variables that need to be controlled. Table 8.2 identifies some of these and gives the reason why the control is necessary.

▼ **Table 8.2 Variables to control in transpiration investigations**

Factor to control	Reason
The size of the leaves (all the same)	Different-sized leaves would have different surface areas
The type of leaf (all from the same species)	Different species may have different numbers of stomata or different thicknesses of waxy cuticle
Time the leaves are left (all left for the same time)	The longer they are left, the more water may be lost
Amount of petroleum jelly smeared on the leaf surface (same amount for each surface treated)	More jelly would give more protection from water loss
Temperature (same for each leaf)	Transpiration is dependent on temperature
Air currents (each leaf exposed to the same amount of wind movement)	Transpiration is dependent on wind speed

Translocation REVISED

Translocation is the movement of sugars and amino acids in the phloem from sources to sinks. The **source** is where the materials are released (usually the leaves) and the **sink** is the region they are transported to. This may be for storage (e.g. in the roots or developing parts of the plants – new leaves, fruits and seeds), or used for respiration or growth.

During the life of a plant, a region that originally acted as a sink may become a source. For example, sugars stored in the leaves of a bulb in the summer (acting as sink) may be translocated to a growing flower bud or stem the following spring. The bulb is now acting as a source.

Revision activity

Try identifying the variables that need to be controlled in other experiments and investigations that you are familiar with.

Exam-style questions

1 Study the diagram of a root hair cell in Figure 8.4.
  a Which plant cell part is missing from this cell? [1]
  b Name the process by which the cell absorbs:
    i water [1]
    ii mineral ions [1]

2 Study the experiment set up in Figure 8.8.
Complete the table to show the results you predict and to give
your reasons. [8]

Leaf	Expected result	Reason
A		
B		
C		
D		

3 Figure 8.9 shows part of the lower surface of a typical dicotyledonous
leaf.

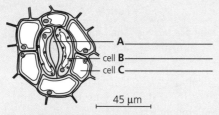

45 µm

▲ **Figure 8.9**

a On the figure, label part A and the cells B and C. [3]
The surfaces of the leaves of two species of plant were studied
and the number of stomata per unit area (stomatal frequency) was
recorded.
Cobalt chloride paper changes colour in the presence of water.
Pieces of cobalt chloride paper were attached to the upper and lower
surfaces of leaves on both plants. The plants were set up for 1 hour
during the day. Any colour changes were recorded. The experiment
was repeated for 1 hour at night. The table shows the results.

| Plant species | Stomatal frequency | | Colour change to cobalt chloride paper | | | |
| | | | Day | | Night | |
	Lower surface	Upper surface	Lower surface	Upper surface	Lower surface	Upper surface
Cassia fistula	0	18	✗	✓	✗	✗
Bauhinia monandra	22	0	✓	✗	✗	✗
Key: ✓ colour change; ✗ no colour change						

b Describe the differences in stomatal distribution between the
two species of plant. [2]
c i Explain the colour changes to the cobalt chloride paper
during the day. [3]
ii Suggest why there was no colour change for either plant at
night. [1]
d Outline the mechanism by which water in the roots reaches the
leaf. [3]
e State and explain the effect of the following on transpiration
rate:
i increasing humidity [2]
ii increasing temperature [2]

Key objectives

The objectives for this chapter are to revise:
- the circulatory system, including the main blood vessels
- the structure of the mammalian heart
- ways of monitoring heart activity
- investigations into the effect of physical activity on pulse rate
- coronary heart disease and its possible risk factors
- structures and functions of arteries, veins and capillaries
- components of blood – red and white blood cells, platelets and plasma – and their functions
- the roles of blood clotting

- single and double circulatory systems
- functioning of the heart in relation to structure
- the effect of physical activity on heart rate
- how the structures of blood vessels are related to their functions
- the functions of lymphocytes and phagocytes
- the process of blood clotting

Key terms

REVISED ☐

Term	Definition
Circulatory system	A system of blood vessels with a pump and valves to ensure one-way flow of blood

Circulatory systems

REVISED ☐

Circulatory systems can be single (e.g. in fish) or double (e.g. in mammals).

Single circulation of fish

Fish have a heart that consists of one blood-collecting chamber (the atrium) and one blood ejection chamber (the ventricle). It sends blood to the gills, where it is oxygenated. The blood then flows to all parts of the body before returning to the heart (Figure 9.1). This is known as single circulation because the blood goes through the heart once for each complete circulation of the body. Pressure is lost as the blood passes through the capillaries of the gills and is not built back up again until the blood returns to the heart. This makes single circulation inefficient.

Gills

Body

▲ **Figure 9.1 Single circulation**

Double circulation of mammals

Blood passes through the heart twice for each complete circulation of the body. The right side of the heart collects deoxygenated blood from the body and pumps it to the lungs. The left side collects oxygenated blood from the lungs and pumps it to the body. The double circulatory system helps to maintain blood pressure, making circulation efficient. Figure 9.2 shows the double circulatory system.

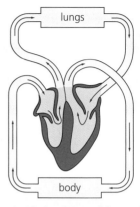

▲ **Figure 9.2 Double circulation**

Heart

The mammalian heart is a pump made of muscle that moves blood around the body. The muscle is constantly active, so it needs its own blood supply – the coronary artery – to provide it with oxygen and glucose (it does not take oxygen or glucose from the blood passing through the chambers of the heart).

Veins carry blood to the heart, while arteries carry blood away from the heart.

The heart has two sides, separated by the septum. The right side receives deoxygenated blood from the body and then pumps it to the lungs for oxygenation. The left side receives oxygenated blood from the lungs and pumps it to the body.

Note that on diagrams of the heart, the right chambers are on the left of the diagram.

There are four chambers. The right and left atria receive blood from veins and squeeze it into the ventricles. The right and left ventricles receive blood from the atria and squeeze it into the arteries. Figure 9.3 shows the main parts of the heart. A surface view of the heart would also show the coronary arteries on the surface of the ventricle muscle walls.

For the core paper, the valves in the heart do not need to be named. You only need to be able to identify where the valves are and understand that they allow the flow of blood only one way.

> **Revision activity**
>
> Create mnemonics to help you remember details about the heart, for example:
> - LORD: **L**eft **O**xygenated **R**ight **D**eoxygenated
> - AA: **A**rteries carry blood **A**way from the heart.

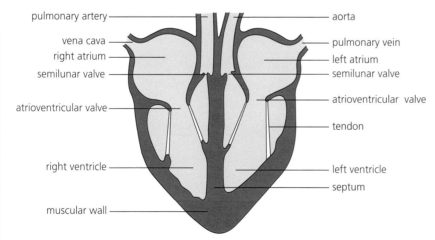

▲ **Figure 9.3 Structure of the heart**

You need to be able to identify the atrioventricular and semilunar valves (labelled on Figure 9.3).

The wall of the left ventricle is much thicker than the wall of the right ventricle because it needs to build up enough pressure to pump the blood to all of the main organs. The right ventricle pumps blood only to the lungs, which is a shorter distance.

The walls of the atria are much thinner than those of the ventricles. This is because the contraction of the atria only needs to be powerful enough to move blood down into the ventricles, while the ventricles are moving blood around the body and through all of the organs.

The septum divides the left side of the heart from the right side. This prevents the mixing of oxygenated and deoxygenated blood.

Control of blood flow through the heart

Heart muscles in the atria contract to build up sufficient pressure to move blood through the atrioventricular valves into the ventricles. These valves then shut to prevent the backflow of blood. As the muscles in the ventricles contract, blood pressure builds up and the blood is forced through the semilunar valves into the pulmonary artery (right side) and aorta (left side). Once the pressure wave has passed, the semilunar valves close to prevent blood from the arteries being sucked back into the ventricles.

> **Revision activity**
>
> On a copy of the diagram of the double circulatory system (Figure 9.2), label:
> ● the four main blood vessels
> ● the chambers of the heart
>
> Draw in the two semilunar valves.

Monitoring the activity of the heart

There are different ways in which the activity of the heart can be monitored:

- **Pulse rate** – the ripple of pressure that passes down an artery as a result of the heartbeat can be felt as a 'pulse' when the artery is near the surface of the body.
- **Heart sounds** can be heard using a **stethoscope**. This instrument amplifies the sounds of the heart valves opening and closing.
- **ECG (electrocardiogram)** – to obtain an ECG, electrodes, attached to an ECG recording machine, are stuck onto the surface of the skin on the arms, legs and chest. Electrical activity associated with the heartbeat is then monitored and viewed on a computer screen or printed out.

The effect of physical activity on heart rate

A heartbeat is a contraction, each of which forces blood to the lungs and body. The heart beats about 70 times a minute – more if you are younger – and the rate becomes lower the fitter you are. This beat can be felt as a pulse in the wrist (radial artery) or neck (carotid artery). During exercise, the heart rate increases from the resting rate and stays high until physical activity slows down or stops. After exercise, the heart rate gradually returns to normal.

> **Revision activity**
>
> Trace the diagram of a heart in Figure 9.3. Then:
> ● Draw arrows in the four blood vessels and the four chambers to show the directions of blood flow through the heart.
> ● Shade the right chambers blue to show deoxygenated blood.
> ● Shade the left chambers red to show oxygenated blood.

> During exercise, the heart rate increases to supply the muscles with more oxygen and glucose. These are needed to allow the muscles to respire aerobically, so they have sufficient energy to contract.

Coronary heart disease

Coronary heart disease is caused by blockage of the coronary arteries, which supply the heart muscle with oxygen and glucose. Without these, the muscle cells stop contracting and die (a heart attack). The possible risk factors are shown in Table 9.1.

▼ Table 9.1 Risk factors for coronary heart disease

Risk factor	Explanation
Poor diet with too much saturated (animal) fat	Leads to fatty deposits (atheroma) in arteries, which eventually block the blood vessel or allow a blood clot to form
Obesity	Being overweight puts extra strain on the heart and makes it more difficult for the person to exercise
Smoking	Nicotine damages the heart and blood vessels
Stress	Tends to increase blood pressure, which can result in fatty materials collecting in the arteries
Lack of exercise	The heart muscle loses its tone and becomes less efficient in pumping blood
Genetic predisposition	Heart disease appears to be passed from one generation to the next in some families
Age	Risk increases with age
Sex	Males are more at risk than females

Prevention of coronary heart disease

Maintaining a healthy, balanced diet will lower the chance of a person becoming obese. The low intake of saturated fats that is part of a balanced diet reduces the chances of a build-up of fatty deposits and the formation of blood clots.

Regular, vigorous exercise can also reduce the chances of a heart attack. This may be because it increases muscle tone – not only of skeletal muscle, but also of cardiac muscle. Good heart muscle tone leads to an improved coronary blood flow and the heart requires less effort to keep pumping. An obese person is less likely to take regular exercise.

Sample question

REVISED

Figure 9.4 shows a section through the heart.

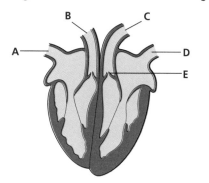

a Name the two blood vessels A and B. [2]
b Which of the blood vessels A, B, C and D carry oxygenated blood? [1]
c Name structure E and state its function. [3]

▲ Figure 9.4

Student's answer

a A, vena cava ✓; B, pulmonary vein ✗
b C ✗
c Name, valve ✓; function, to stop blood going backwards ✓.

Teacher's comments

Blood vessel B is the pulmonary artery. Arteries of the heart always carry blood from a ventricle. Part b needs two answers (blood vessels C and D) to gain the mark. Blood vessel D is the pulmonary vein, which carries oxygenated blood to the heart from the lungs. Blood vessel C is the aorta, which carries oxygenated blood from the heart to the body. In part c, 'valve' is correct, but there are 2 marks for its function. This student has given only one statement – a second mark was available for stating that the valve prevents blood from going back *into the left ventricle*.

Blood vessels

REVISED ☐

Arteries carry blood, at high pressure, away from the heart to the organs of the body. **Veins** return blood, at low pressure, from the organs towards the heart. **Capillaries** link arteries to veins. They carry blood through organs and tissues, allowing materials to be exchanged.

Table 9.2 compares the structures of arteries, veins and capillaries. An explanation of how the structure is related to the function is needed only for the extended paper.

▼ **Table 9.2 Structure and function of arteries, veins and capillaries**

Blood vessel	Structure	How structure is related to function
Artery	Thick, tough wall with muscles, elastic fibres and fibrous tissue	Carries blood at high pressure – prevents bursting and maintains pressure wave
	Lumen quite narrow, but increases as a pulse of blood passes through	This helps to maintain blood pressure
	Valves absent	The high pressure prevents blood flowing backwards, so valves are not necessary
Vein	Thin wall of mainly fibrous tissue, with little muscle and few elastic fibres	Carries blood at low pressure
	Lumen large	This reduces resistance to blood flow
	Valves present	To prevent backflow of blood
Capillary	Permeable wall, one cell thick, with no muscle or elastic tissue	Allows diffusion of materials between capillary and surrounding tissues
		White blood cells can squeeze between cells of the wall
	Lumen approximately one red blood cell wide	Red blood cells pass through slowly to allow diffusion of materials and tissue fluid
	Valves absent	Blood is still under pressure

Figure 9.3 shows all the blood vessels associated with the heart and lungs that you need to learn. Blood vessels associated with the kidneys (the renal artery and renal vein) are shown in Figure 13.1 (p. 80).

> The liver is served by the hepatic artery. It also receives blood through the hepatic portal vein, bringing products of digestion from the intestine. Blood leaves the liver via the hepatic vein.

Blood

Blood is made up of a liquid (plasma) containing red and white blood cells, and platelets. Figure 9.5 gives details of the components of blood and their functions. The terms used for naming white blood cells (lymphocytes and phagocytes) are needed only for the extended paper.

red blood cell

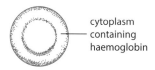

cytoplasm containing haemoglobin

biconcave discs with no nucleus – carry oxygen

lymphocyte

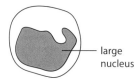

large nucleus

produce antibodies to fight bacteria and foreign materials

phagocyte

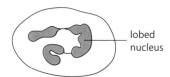

lobed nucleus

fight disease by surrounding bacteria and engulfing them – a process called phagocytosis

platelets

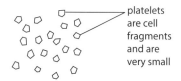

platelets are cell fragments and are very small

form blood clots, which stop blood loss at a wound and prevent the entry of pathogens into the body

▲ **Figure 9.5 Components of blood and their functions**

Skills

Drawing skills

Drawing skills are often tested in the Practical and Alternative to Practical examinations. For example, you may be given a photomicrograph of a blood cell to draw. The following points will help you to gain maximum marks:
- Use a sharp HB pencil.
- Draw only what you are told to draw – for example, one cell.
- Make your drawing large (at least one quarter of an A4 sheet of paper).
- Draw the outline first, to get the dimensions of the image correct, and then add details you can see.
- Use clean, unbroken lines.
- Avoid shading or using colour.
- Only add labels if told to do so.
- Write the labels horizontally outside the drawing.
- Link each label to the part with a ruled pencil line, touching the part it is labelling, while making sure that the label lines do not cross each other.
- Give the drawing a title (in this case, the type of cell).
- If told to do so, add a magnification to the drawing. This will be an estimation, based on how many times bigger than the image your drawing is, multiplied by the photomicrograph magnification.

Plasma is a liquid that transports substances to cells and carries waste away from cells. It acts as a pool for amino acids (these cannot be stored in the body) and contains blood proteins that are important in blood clotting. Table 9.3 shows the main substances carried by plasma.

▼ **Table 9.3 Substances carried by plasma**

Substance carried in plasma	From	To
Amino acids	Small intestine	Sites of growth and repair
Carbon dioxide	Respiring tissues	Lungs
Glucose	Small intestine	All tissues
Heat	Liver, muscles	All tissues
Hormones, e.g. insulin	Endocrine glands, e.g. pancreas	Target organs, e.g. liver
Urea	Liver	Kidneys
Ions	Small intestines, kidneys	All tissues

Transport of oxygen

Oxygen is not included in Table 9.3 because it is transported in red blood cells. Oxygen combines with haemoglobin to form oxyhaemoglobin. The oxygen is released from the red blood cells in capillaries where surrounding oxygen levels are lower.

Sample question

a State two structures that red blood cells and white blood cells have in common. [2]

b Suggest why it is an advantage for red blood cells to have no nucleus. [2]

c White blood cells help to protect against infection by bacteria. Some white blood cells produce antibodies. Describe how other white blood cells fight bacteria. [1]

Student's answer

> a Cytoplasm ✓ and cell wall ✗
> b There is more cytoplasm, which contains haemoglobin ✓ to carry oxygen ✓.
> c They kill bacteria. ✗

Teacher's comments

In part a, the student has forgotten that blood cells are animal cells, which do not have a cell wall. The cytoplasm is surrounded by a cell membrane.

The answer to part b is very good, gaining both available marks. This is a 'suggest' question, so you are not expected to know the answer, but to make a sensible suggestion, based on your knowledge of biology, to explain the lack of nucleus in red blood cells. It is always worth writing an answer even if you are not sure, because you will not be penalised if your answer is incorrect.

Part c does not gain a mark because the response is too vague. It does not state how they kill the bacteria. A mark-worthy answer would include 'by phagocytosis', or state 'they engulf the bacteria' or 'they digest bacteria'.

Functions of lymphocytes and phagocytes

Lymphocytes are involved in the production of antibodies, which are needed to fight disease (see Chapter 10). Antibodies can attach themselves to antigens (foreign proteins) and clump them together.

Phagocytes have the ability to change their shape and move to engulf pathogens by a process called **phagocytosis**.

Clotting

Platelets clump together when tissues are damaged and block the smaller capillaries. The platelets and damaged cells at the wound produce a substance that acts on a soluble plasma protein called **fibrinogen**. As a result, it is changed into insoluble **fibrin**, which forms a network of fibres across the wound. Red blood cells become trapped in this network and so form a blood clot. The clot not only stops further loss of blood, but also prevents the entry of **pathogens** (disease-causing organisms) into the wound.

Exam-style questions

1 a State two similarities and two differences between a single
      circulation and a double circulation.   [4]
   b Explain the advantages of a double circulation.   [2]

2 Name the parts associated with the heart that:
   a prevent blood flowing backwards   [1]
   b receive blood from the veins   [1]
   c transfer blood to the lungs   [1]
   d provide the heart muscle with oxygen   [1]
   e are chambers containing oxygenated blood   [2]
3 a State three risk factors for coronary heart disease, other than
      poor diet and lack of exercise.   [3]
   b Discuss the roles of diet and exercise in reducing the risk of
      coronary heart disease.   [4]

4 Figure 9.6 shows the relationship between the heart, the liver, the
   small intestine and the main blood vessels linking them.

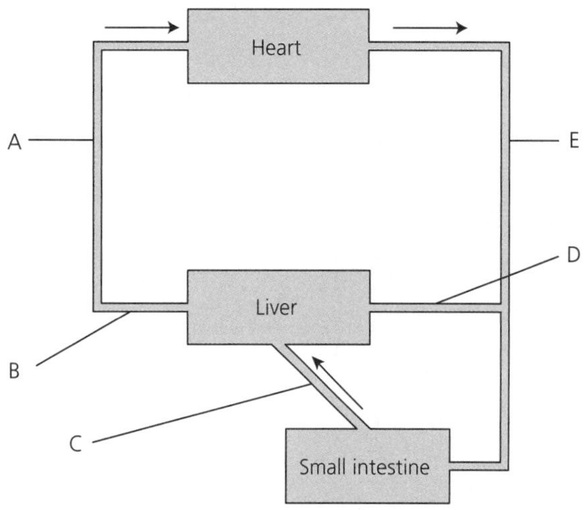

▲ **Figure 9.6**

   Name the blood vessels labelled A–E.   [5]

5 a State two differences, other than colour, between red blood
      cells and white blood cells.   [2]
   b i   State which component of the blood is involved in blood
          clotting.   [1]
      ii State two roles of blood clotting.   [2]

Diseases and immunity

Key objectives

The objectives for this chapter are to revise:
- definitions of the key terms
- how the pathogen for a transmissible disease may be transmitted
- the defences of the body and ways of controlling the spread of disease

- how antibodies work and how to explain their specificity
- how active immunity is gained and the process and role of vaccination

- the characteristics and importance of passive immunity
- the importance of breastfeeding for the development of passive immunity in infants
- the cause of cholera and how it is transmitted
- how the cholera bacterium causes diarrhoea

Key terms

REVISED

Term	Definition
Pathogen	A disease-causing organism
Transmissible disease	A disease in which the pathogen can be passed from one host to another
Active immunity	Defence against a pathogen by antibody production in the body
Antibodies	Proteins that bind to antigens, leading to direct destruction of pathogens or marking of pathogens for destruction by phagocytes

Pathogens and transmission

REVISED

Pathogens responsible for **transmissible diseases** can be spread either through direct contact or indirectly. When writing about disease-causing organisms, remember to use the term 'pathogens' rather than 'germs'.

Direct contact

Direct contact may involve transfer through blood or other body fluids.

- For example, HIV can be passed on by drug addicts who inject drugs into their bloodstream if they share needles with other drug users, as the needle will be contaminated.
- Anyone cleaning up dirty needles is at risk of infection if they accidently stab themselves.
- Surgeons carrying out operations have to be especially careful not to be in direct contact with the patient's blood.
- A person with HIV or another sexually transmitted disease (see Chapter 16) who has unprotected sex can pass on the pathogen to their partner through body fluids.

Indirect contact

Indirect contact can involve infection from pathogens on contaminated surfaces, for example during food preparation.

- Raw meat carries bacteria that are killed if the meat is cooked adequately. However, if the raw meat is prepared on a surface that is then used for other food preparation, for example cutting up vegetables that are later eaten raw, the pathogens from the meat can be transferred to the fresh food.
- People handling food are also potential vectors of disease, for example if they do not wash their hands after using the toilet.
- Intensive methods of animal rearing may contribute to the spread of infection unless care is taken to reduce the exposure of animals to infected faeces.
- Houseflies act as animal vectors. They walk about on exposed food and place their mouthparts on it. Then they pump saliva onto the food and suck up the digested food as a liquid. Unfortunately, they may have already carried out this process on contaminated food or faeces, transferring bacteria in the process.
- Airborne infections can be spread by a person with an infection sneezing or coughing. Droplets containing the pathogen float in the air and may be breathed in by other people or fall onto exposed food. Examples of diseases spread in this way include colds, flu, measles and sore throats.

Pathogens cannot usually be passed on by touching a person with the disease. The pathogen is carried in body fluids such as blood. However, food can be contaminated when a person with pathogens on their skin (e.g. dirty hands) handles it.

Controlling disease

- Clean water supply: faeces contain bacteria, which can contaminate water used for drinking, such as streams, rivers and reservoirs. On a small scale, drinking water can be boiled to destroy any pathogens. On a large scale, drinking water can be protected from sewage contamination, or treated to make it safe. This treatment involves filtration and chlorination.
- Hygienic food preparation: keeping food-preparation surfaces clean, avoiding the preparation of raw and cooked food on the same surface, cooking food thoroughly to kill any bacteria present.
- Good personal hygiene: washing hands after using the toilet, moving rubbish or handling raw food; avoiding the handling of money when in contact with unwrapped food.
- Waste disposal: to avoid the development of a breeding ground for pathogens.
- Sewage treatment: to prevent pathogens in faeces from contaminating drinking water and to stop vectors such as flies or rats feeding and transmitting the disease organism.

Cholera

This disease is caused by the bacterium *Vibrio cholerae*, which results in acute diarrhoea. It is transmitted in contaminated water – for example, from sewage leaking into the drinking water supply.

How cholera causes diarrhoea

Figure 10.1 summarises how cholera causes diarrhoea. When *Vibrio cholerae* bacteria are ingested, they multiply in the small intestine and invade its epithelial cells. As the bacteria become embedded, they release toxins (poisons) that irritate the intestinal lining and lead to the secretion of large amounts of water and salts, including chloride ions. The salts decrease the osmotic potential of the gut contents, drawing more water from surrounding tissues and blood by osmosis (see Chapter 3). This makes the undigested food much more watery, leading to acute diarrhoea, while the loss of body fluids and salt leads to dehydration and kidney failure.

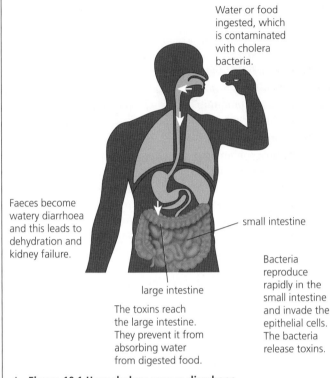

Water or food ingested, which is contaminated with cholera bacteria.

Faeces become watery diarrhoea and this leads to dehydration and kidney failure.

small intestine

Bacteria reproduce rapidly in the small intestine and invade the epithelial cells. The bacteria release toxins.

large intestine

The toxins reach the large intestine. They prevent it from absorbing water from digested food.

▲ **Figure 10.1 How cholera causes diarrhoea**

Diarrhoea

Diarrhoea is the loss of watery faeces. It is sometimes caused by bacterial or viral infection, for example from food or water, resulting in the intestines being unable to absorb fluid from the contents of the colon or too much fluid being secreted into the colon. Undigested food then moves through the large intestine too quickly, resulting in insufficient time to absorb water from it. Unless the condition is treated, **dehydration** can occur.

The treatment is called **oral rehydration therapy** – drinking plenty of fluids (sipping small amounts of water at a time) to rehydrate the body.

Defences against diseases

REVISED

The body has three main lines of defence against disease:

- Mechanical barriers:
  - The epidermis of the skin is a barrier that prevents bacteria getting into the body.
  - Hairs in the nose help to filter out bacteria that are breathed in.
- Chemical barriers:
  - The acid conditions created by hydrochloric acid in the stomach destroy most of the bacteria that might be taken in with food.
  - Mucus, produced by the lining of the trachea and bronchi, is sticky and traps pathogens.

- Cells:
  - One type of white blood cell produces **antibodies** that attach themselves to bacteria, making it easier for other white blood cells to engulf them.
  - Another type of white blood cell engulfs bacteria (a process called phagocytosis) and digests them (see Chapter 9).

Antibodies and immunity

Antibodies are proteins that bind to **antigens**. They are produced by lymphocytes, formed in lymph nodes in response to the presence of pathogens such as bacteria. The pathogens have antigens on their surface; there is a different antigen with a specific shape for each type of pathogen. Therefore, a different antibody with a complementary shape has to be produced for each antigen. The antibodies make bacteria clump together and mark them in preparation for destruction by phagocytes, or neutralise the toxins produced by the bacteria. Once antibodies have been made, they remain in the blood to provide long-term protection.

Active immunity

Some of the lymphocytes that produced the specific antibodies, as a result of infection by a pathogen, remain in the lymph nodes as **memory cells** for some time, and can divide rapidly to produce more antibodies to respond to further infections by the same pathogen. This creates **immunity** to the disease caused by the antigen.

Active immunity can also be achieved by vaccination.

Vaccination

- Vaccination gives a person immunity to a specific disease organism that may otherwise be life threatening if a person is infected by it.
- Vaccination involves a weakened form of the pathogen that has antigens, or antigens from the pathogen, being introduced into the body by injection or swallowing.
- The presence of the antigens triggers an immune response: lymphocytes make specific antibodies to combat possible infection.
- Some of these cells remain in the lymph nodes as memory cells.
- These can reproduce quickly and produce antibodies in response to any subsequent invasion of the body by the same pathogen, providing long-term immunity.
- Mass vaccination can control the spread of diseases. A significant proportion of a population needs to be immunised to prevent an epidemic of a disease – ideally over 90%. If mass vaccination fails, the population is at risk of infection, with the potential for epidemics.

Passive immunity

- Passive immunity is a short-term defence against a pathogen.
- It is achieved by injecting the patient with serum taken from a person who has recovered from the disease. Serum is plasma with the fibrinogen removed. It contains antibodies against the disease – for example, tetanus, chickenpox, rabies.
- It is called passive immunity because the antibodies have not been produced by the patient. It is only temporary because it does not result in the formation of memory cells.

- When a mother breastfeeds her baby, the milk contains some of the mother's lymphocytes, which produce antibodies.
- These antibodies provide the baby with protection against infection at a time when the baby's immune responses are not yet fully developed. However, this is another case of passive immunity, as it is only short-term protection – memory cells are not produced.

Sample question

REVISED

Distinguish between the terms *active immunity* and *passive immunity*. [4]

Student's answer

Active immunity is a defence against pathogens ✓ by producing antibodies to fight them ✓. Passive immunity is the same ✗, but is only short term ✗.

Teacher's comments

The definition of active immunity is correct. However, passive immunity is not the same as active immunity; in passive immunity, the antibodies are acquired from another individual. The statement about passive immunity being short term is correct, but is given as part of a biologically incorrect statement, so it does not gain a mark.

Correct answer

Active immunity can be gained by infection with a pathogen, or by vaccination. In active immunity the body produces antibodies to respond to further infections by the same pathogen. In passive immunity, antibodies are acquired from another individual. Passive immunity only offers short-term protection.

Revision activity

Make a set of flashcards for this topic:
- You could include details such as the key definitions, how a pathogen is transmitted, and a list of body defences.
- Look at each flash card in turn, put it out of sight, and then try to say what was on the card, or write it down.
- Repeat the process until you manage to get it right. Then move on to the next card. Keep the set of cards for revision before your exams.

Exam-style questions

1 a Define the term *transmissible disease*. [1]
  b State two ways in which pathogens can be transmitted:
    i directly [2]
    ii indirectly [2]

2 Describe the role of osmosis in causing diarrhoea in a person infected with cholera bacteria. [4]
3 a Draw a diagram to show how lymphocytes and phagocytes fight pathogens and destroy them. Annotate your diagram. [6]
  b Distinguish between active and passive immunity. [4]

11 Gas exchange in humans

Key objectives

The objectives for this chapter are to revise:
- the features of gaseous exchange surfaces in humans
- structures associated with the breathing system
- differences in composition between inspired and expired air, and the test for carbon dioxide
- the effects of physical activity on the rate and depth of breathing
- the functions of cartilage in the trachea
- the role of the ribs, intercostal muscles and diaphragm in ventilation of the lungs
- how to explain differences in composition between inspired and expired air
- how to explain the link between physical activity and the rate and depth of breathing
- how to explain the role of goblet cells, mucus and ciliated cells in protecting the breathing system

Gas exchange in humans

REVISED

This process involves the passage of gases such as oxygen into and carbon dioxide out of cells or a transport system. First, air needs to be in contact with the gaseous exchange surface. This is achieved by breathing. Figure 11.1 shows the breathing system of a human.

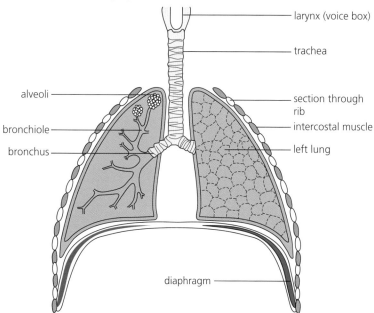

▲ **Figure 11.1 The human breathing system**

Revision activity

To help you remember the sequence of structures through which air passes when you breathe in, imagine an apple tree:
- Trunk   Trachea
- Bough   Bronchus
- Branches   Bronchioles
- Apples   Alveoli

Sample question

REVISED

A new, alternative treatment for diabetes is being developed that involves inhaling insulin into the lungs as a spray.

Suggest the path the spray would take from the mouth to enter the alveoli. [3]

Student's answer

> The spray would pass through the trachea, then through the bronchioles, then the bronchi to the alveoli. ✓✓

Teacher's comments

The student has named all the parts involved, but bronchioles and bronchi are the wrong way round, so only 2 marks are awarded.

Features of respiratory surfaces

Gaseous exchange relies on diffusion. To be efficient, the gaseous exchange surface must:

- be thin – a short distance for gases to diffuse
- have a large surface area – for gases to diffuse over
- have good ventilation with air – this creates and maintains a concentration gradient
- have a good blood supply – to transport oxygen to respiring tissues and bring carbon dioxide from those tissues

The gaseous exchange surfaces in mammals are the alveoli in the lungs. Figure 11.2 shows the features supporting gaseous exchange in an alveolus. Figure 11.3 shows the blood supply of the alveoli.

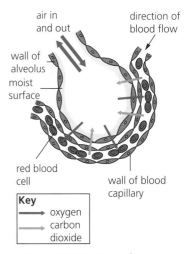

▲ Figure 11.2 Features for gaseous exchange in an alveolus

The composition of inspired and expired air

You need to be able to state the percentages shown in Table 11.1. However, the explanations are needed only for the extended paper.

▼ Table 11.1 Composition of inspired and expired air

Gas	Inspired air/%	Expired air/%	Explanation
Nitrogen	79	79	Not used or produced by body processes
Oxygen	21	16	Used up in the process of respiration, but the system is not very efficient, so only a small proportion of the oxygen available is absorbed from the air
Carbon dioxide	0.04	4	Produced in the process of respiration
Water vapour	Variable	Saturated	Produced in the process of respiration; moisture evaporates from the surface of the alveoli

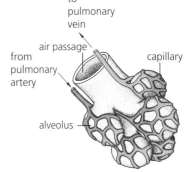

▲ Figure 11.3 Blood supply of the alveoli

Testing for carbon dioxide

Limewater can be used to test for carbon dioxide – it changes colour from colourless to milky when the gas is bubbled through.

You need to be able to describe how to investigate the differences in composition between inspired and expired air, using limewater as a test for carbon dioxide. Using the apparatus in Figure 11.4 is one way of doing this.

When you breathe in through the apparatus, air is sucked in through boiling tube A. When you breathe out, air is blown through boiling tube B. The limewater reacts with any carbon dioxide in the air bubbled through it and changes from colourless to milky. Expired air makes limewater change colour more quickly than inspired air because there is more carbon dioxide present in expired air.

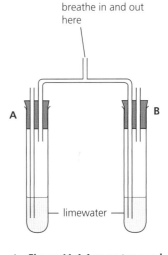

▲ Figure 11.4 Apparatus used to test for carbon dioxide

Skills

Calculating percentage change

The formula for percentage change is:

$$\frac{\text{change}}{\text{starting value}} \times 100$$

For example, in Table 11.1 the percentage of oxygen in inspired air is 21%. This falls to 16% in expired air. Therefore, the change is 21 – 16 = –5%.

Applying these values to the equation, the percentage change in oxygen between breathing in and out is:

$$\frac{-5}{21} \times 100 = -23.8\%$$

Effects of physical activity on breathing

Both breathing rate and depth increase during exercise. The volume of air breathed in and out during normal, relaxed breathing is about 0.5 litres. This is the **tidal volume**. Breathing rate at rest is about 12 breaths per minute.

During exercise, the volume inhaled (depth) increases to about 5 litres (depending on the age, sex, size and fitness of the person). The maximum amount of air breathed in or out in one breath is the **vital capacity**. Breathing rate can increase to over 20 breaths per minute.

For limbs to move faster during exercise, aerobic respiration in the skeletal muscles increases. Carbon dioxide is a waste product of aerobic respiration (see Chapter 12). As a result, carbon dioxide builds up in the muscle cells and diffuses into the plasma in the bloodstream more rapidly. The brain detects increases in carbon dioxide concentration in the blood and stimulates the breathing mechanism to speed up, increasing the rate of expiration of the gas.

You need to be able to describe how to investigate the effects of physical activity on the rate and depth of breathing. This can be done with an instrument called a spirometer. It records, digitally or on paper, the volume of air being inspired and expired by a person breathing through the apparatus. From the data produced, the volume of air inspired and expired per breath can be calculated. The number of breaths per minute and the volume of oxygen used by the person can also be worked out.

Ventilation of the lungs

Figure 11.5 shows the relationship between the intercostal muscles, diaphragm and ribcage that is required to achieve ventilation of the lungs.

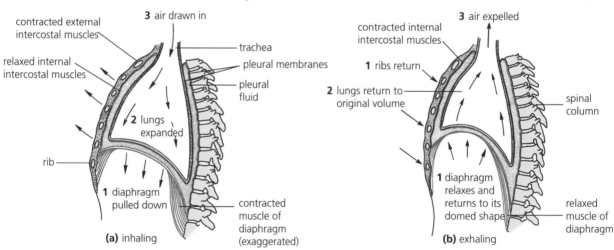

▲ **Figure 11.5 The intercostal muscles, diaphragm and ribcage**

The movement of the ribcage is brought about by the contraction of two sets of intercostal muscles, which are attached to the ribs. The external intercostal muscles are attached to the external surface of the ribs; the internal intercostal muscles are attached to the internal surface.

You need to be able to identify the external and internal intercostal muscles in images and diagrams.

The diaphragm is a tough, fibrous sheet at the base of the thorax, with muscle around its edge.

Inspiration (breathing in)

- When the external intercostal muscles contract, they move the ribcage upwards and outwards, increasing the volume of the thorax.
- When the diaphragm muscle contracts, the diaphragm moves down, again increasing the volume of the thorax.
- This increase in volume reduces the air pressure in the thoracic cavity.
- As the air pressure outside the body is higher, air rushes into the lungs through the mouth or nose.

Expiration (breathing out)

- The opposite of inspiration happens when breathing out.
- During forced exhalation, the internal intercostal muscles contract and the diaphragm muscles relax, pulling the ribs downwards.
- Thoracic volume decreases, so air pressure becomes greater than outside the body.
- Air rushes out of the lungs to equalise the pressure.

Protection of the gas exchange system from pathogens and particles

Pathogens (e.g. bacteria) and dust particles are present in the air we breathe in, and are potentially dangerous if not actively removed. Two types of cell provide mechanisms to help achieve this (Figure 11.6).

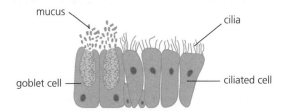

▲ **Figure 11.6 Epithelial lining of the respiratory tract**

- **Goblet cells** are found in the epithelial lining of the trachea, bronchi and some bronchioles of the respiratory tract. Their role is to secrete **mucus**. The mucus forms a thin film over the internal lining. This sticky liquid traps pathogens and small particles, preventing them from entering the alveoli, where they could cause infection or physical damage.
- **Ciliated cells** (Figure 2.5, p. 14) are also present in the epithelial lining of the respiratory tract. Their continual flicking motion moves the mucus, secreted by the goblet cells, upwards and away from the lungs. When the mucus reaches the top of the trachea, it passes down the gullet during normal swallowing.

Exam-style questions

1   State how each feature labelled on the diagram of an alveolus (Figure 11.2) makes the process of gaseous exchange efficient.   [5]
2   Calculate the percentage change in volume of carbon dioxide between inspired air and expired air.   [2]
3   a   The composition of the air inside the lungs changes during breathing.
    i   State three differences between inspired and expired air.   [3]
    ii   Gaseous exchange in the alveoli causes some of the changes to the inspired air. Describe three features of the alveoli that assist gaseous exchange.   [3]
  b   i   State what is meant by anaerobic respiration.   [2]
    ii   Where does anaerobic respiration occur in humans?   [1]

4   Write out the series of events involved in breathing in and breathing out as a set of bullet points or as a flowchart linked with arrows.   [6]

12 Respiration

Key objectives

The objectives for this chapter are to revise:
- definitions of the key terms
- the uses of energy in living organisms
- investigations into the effect of temperature on respiration in yeast
- the word equations for aerobic and anaerobic respiration in muscles
- that anaerobic respiration releases far less energy per glucose molecule than aerobic respiration

- the balanced chemical equations for aerobic respiration and for anaerobic respiration in yeast
- the way in which an oxygen debt builds up and how it is removed during recovery

Key terms

REVISED

Term	Definition
Aerobic respiration	The chemical reactions in cells that use oxygen to break down nutrient molecules to release energy
Anaerobic respiration	The chemical reactions in cells that break down nutrient molecules to release energy without using oxygen

Respiration

REVISED

Most of the processes taking place in cells in the body need energy to make them happen. Examples of energy-consuming processes in living organisms are:

- the contraction of muscle cells – for example, to create movement of the organism
- synthesis (building up) of proteins from amino acids
- the process of cell division (Chapter 17) to create more cells, to replace damaged or worn-out cells, or to make reproductive cells
- the process of active transport (Chapter 3), involving the movement of molecules across a cell membrane against a concentration gradient
- growth of an organism through the formation of new cells or a permanent increase in cell size
- the conduction of electrical impulses by nerve cells (Chapter 14)
- maintaining a constant body temperature in warm-blooded animals (Chapter 14) to make sure that vital chemical reactions continue at a steady rate, even when the surrounding temperature varies

This energy comes from the food that cells take in. The food mainly used for energy in cells is glucose. **Respiration**, which releases the energy, is a chemical process that takes place in cells and involves the action of enzymes. Do not confuse it with breathing – the physical process of ventilating the lungs to obtain oxygen and remove carbon dioxide.

The effect of temperature on yeast respiration

You need to be able to describe how to investigate the effect of temperature on respiration in yeast. One investigation uses the apparatus shown in Figure 12.1.

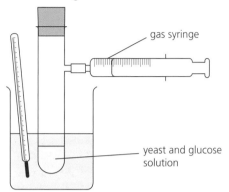

gas syringe

yeast and glucose solution

▲ **Figure 12.1 Investigating the effect of temperature on respiration in yeast**

The water bath is used to control the temperature and the gas syringe collects the carbon dioxide produced as the yeast respires over a chosen amount of time. The experiment is repeated over a range of temperatures – for example, 10°C, 20°C, 30°C, 40°C, 50°C.

As the temperature increases, the volume of carbon dioxide produced also increases, with an optimum at around 35–40°C. Higher temperatures slow down the rate of gas production because the enzymes involved in respiration start to denature (see Chapter 5).

Aerobic respiration

REVISED ☐

In humans, energy is usually released by **aerobic respiration**. However, the cells must receive plenty of oxygen to maintain this process.

The word equation for aerobic respiration is:

glucose + oxygen → water + carbon dioxide

The breakdown of one glucose molecule releases 2830 kJ of energy.

It is possible to carry out experiments using invertebrates and germinating seeds and measure the oxygen uptake of the organisms: the faster the uptake, the faster the rate of aerobic respiration. Germinating seeds do not need energy for movement, so their respiration rate tends to be lower than that of animals.

> If you are following the extended curriculum, you need to be able to state the balanced chemical equation for aerobic respiration:
>
> $$C_6H_{12}O_6 + 6O_2 \rightarrow 6H_2O + 6CO_2$$
>
> If you write a symbol equation, you must make sure that the formulae are correct and that the equation is balanced.
>
> When carrying out experiments using germinating seeds to investigate the effect of temperature on respiration rate, the rate of oxygen uptake is used to indicate respiration rate. As temperature increases, so does the rate of oxygen uptake and, therefore, the respiration rate. This is because respiration is controlled by enzymes. An increase in temperature increases the kinetic energy of the molecules, so the reaction rate increases.

Anaerobic respiration

REVISED

Anaerobic respiration does not require oxygen. When tissues are respiring very fast, the oxygen supply is not fast enough to cope, so tissues such as muscles start to respire anaerobically instead. However, this is a less efficient process than aerobic respiration, so much less energy is produced.

The breakdown of one glucose molecule by yeast releases only 118 kJ of energy.

The word equation for anaerobic respiration in yeast is:

glucose → ethanol + carbon dioxide

Anaerobic respiration in yeast does not produce water.

The word equation for anaerobic respiration in muscles is:

glucose → lactic acid

Anaerobic respiration in muscles does not produce carbon dioxide or water.

The balanced chemical equation for anaerobic respiration in yeast is:

$$C_6H_{12}O_6 \rightarrow 2C_2H_5OH + 2CO_2$$

Revision activity

If you are studying the extended curriculum, you also need to learn the balanced chemical equations for aerobic respiration and anaerobic respiration in yeast.

When you write these down, make sure you have the same number of carbon, oxygen and hydrogen atoms on both sides of each equation. For example, in the aerobic respiration equation there are 6 carbon, 18 oxygen and 12 hydrogen atoms on each side of the reaction arrow.

Oxygen debt

Muscles respire anaerobically when exercising vigorously because the blood cannot supply enough oxygen to maintain aerobic respiration. However, the formation and build-up of lactic acid in muscles causes cramp (muscle fatigue). An oxygen debt is created because oxygen is needed for aerobic respiration to convert lactic acid back to a harmless chemical (pyruvic acid). This happens in the liver.

At the end of a race, a sprinter has to pant to get sufficient oxygen to the muscles to repay the oxygen debt. Breathing remains deep and fast to supply enough oxygen for the aerobic respiration of lactic acid. The heart rate remains fast to transport lactic acid in the blood from the muscles to the liver.

Long-distance runners judge their pace, not running too fast, to prevent the muscles respiring anaerobically. Muscle cramp would stop the athlete running.

Sample question

REVISED

Explain why the breathing pattern changes after a period of vigorous exercise. [3]

Student's answer

The breathing rate increases ✓ because muscles build up an oxygen debt ✓ when they respire anaerobically ✓. Oxygen is needed to break down the lactic acid produced to prevent muscle fatigue.

Teacher's comments

This is an excellent answer, gaining maximum marks. The last sentence also contains creditworthy statements.

Exam-style questions

1 Make a table to compare aerobic and anaerobic respiration in humans, using the headings shown below. [6]

Type of respiration	Requirement(s)	Product(s)	Comparative amount of energy released
Aerobic			
Anaerobic			

2 a State what gas is produced when yeast respires. [1]

   b Suggest two reasons why yeast may die if left in a solution of glucose in a sealed container for 48 hours. [2]

3 a Suggest why a runner taking part in a long-distance race avoids sprinting until near the end of the race. [3]

   b Explain why runners breathe faster during a run than before they start the run. [2]

13 Excretion in humans

Key objectives

The objectives for this chapter are to revise:
- definitions of the key terms
- the organs responsible for excreting carbon dioxide, urea, excess water and ions
- how to identify the kidneys, ureters, bladder and urethra
- the importance of excretion
- the structure of the kidney and the structure and function of a nephron
- the role of the liver in the assimilation of amino acids

Key terms

Term	Definition
Excretion	Removal of waste products of metabolism and substances in excess of requirements
Deamination	The removal of the nitrogen-containing part of amino acids to form urea

Excretion

A number of organs in the body are involved in **excretion**. These include:

- the lungs, which remove carbon dioxide from the blood (see Chapter 11)
- the kidneys, which filter blood, removing urea, excess water and mineral ions

Note that faeces are not an example of excretion – faeces are mainly undigested material that has passed through the gut, but which has not been made in the body. The only excretory materials in it are bile pigments.

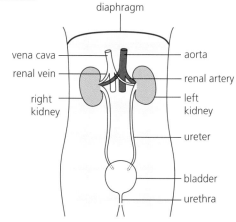

▲ Figure 13.1 The human urinary system

The kidneys

Figure 13.1 shows the relative positions of the kidneys, ureters, bladder and urethra in the body. Filtered blood returns to the vena cava (main vein) via a renal vein. The urine formed in the kidney passes down a ureter into the bladder, where it is stored. A sphincter muscle controls the release of urine through the urethra.

The liquid produced by the filtration of blood in the kidneys is called urine – a solution of urea and mineral ions in water. The relative amount of water reabsorbed depends on the state of hydration of the body (how much water is in the blood).

The need for excretion

Some waste products can be toxic. One example of this is urea.

Urea is not normally harmful unless its concentration in the blood gets too high (e.g. if the kidneys fail to excrete it), in which case it starts to be converted into ammonia – a strong, toxic alkali.

> **Revision activity**
>
> Copy or trace Figure 13.1 and practise labelling it. The spellings of ureter and urethra are really important. Check that you get these right, and that the structures are labelled in the correct positions.

Microscopic structure of the kidneys

Figure 13.2 shows the structure of a kidney, with its related vessels. Figure 13.3 shows the structure of a kidney nephron.

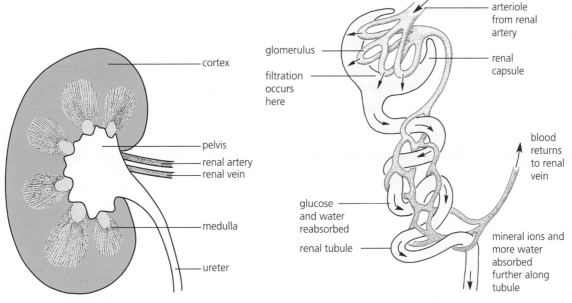

▲ **Figure 13.2 Structure of a kidney**

▲ **Figure 13.3 Structure of a kidney nephron**

The kidneys receive blood from the aorta (the main artery) via the renal arteries. In the cortex, the renal artery splits into millions of capillaries. Each capillary forms a tangled glomerulus, from which the blood is filtered under pressure. This forces all the small molecules and ions (such as glucose, urea, water and mineral ions) out of the capillary into a tubule.

As the filtrate passes through the tubule, reabsorption takes place. Water is reabsorbed by osmosis, while glucose and mineral salts pass back into the blood by diffusion and active uptake (see Chapter 3). The reduction in water content leads to an increase in the concentration of urea. The remaining solution, called urine, passes down the collecting duct of the tubule, through the pelvis of the kidney into the ureter, then down into the bladder for removal through the urethra.

Sample question

REVISED ☐

Outline the function of a renal capsule and renal tubule in the kidney. [4]

Student's answer

After the blood has been filtered, the capsule collects the filtrate. ✓ *This contains water, dissolved salts, glucose and urea.* ✓ *As the liquid passes along the tubule, all the glucose is reabsorbed* ✓*, along with some of the water and salts* ✓*, back into the blood capillaries. The remaining filtrate passes down to the ureter.*

Teacher's comments

This is an excellent answer. Marks could also have been gained by referring to the processes involved in the reabsorption process (diffusion and active transport for the glucose; osmosis for the water).

The liver and its role in dealing with excess amino acids

Surplus amino acids in the bloodstream cannot be stored. They are removed by the liver and some are assimilated by converting them into proteins. Examples of these include plasma proteins (e.g. fibrinogen in the blood; see Chapter 9). The surplus amino acids are broken down into urea (which is the nitrogen-containing part of the amino acid) and a sugar residue (which can be respired to release energy). The breakdown of amino acids is called **deamination**. Urea is returned to the bloodstream (into the hepatic vein) and filtered out when it reaches the kidneys.

Exam-style questions

1 Figure 13.4 shows the water balance of the body. The term *metabolism* refers to chemical processes in cells.

a One of the components of water loss is urine. State two chemicals dissolved in water in urine. [2]

b i Calculate the volume of urine lost, based on the data shown. [2]

ii What percentage of water lost is urine? [2]

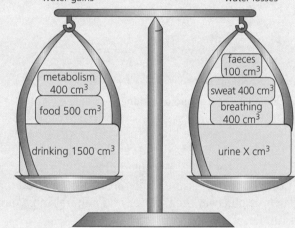

▲ **Figure 13.4**

2 Figure 13.5 shows the human urinary system.

a Name parts X, Y and Z. [3]

b Name the blood vessel that carries blood from the aorta to the kidneys. [1]

c Suggest two differences between the composition of the blood flowing to the kidneys and the blood flowing away from the kidneys. [2]

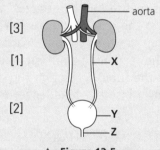

▲ **Figure 13.5**

3 Copy and complete the passage about the function of the kidney, using the words provided. Each word may be used once, more than once, or not at all.

artery	glucose	ureter	vein
capillaries	mineral ions	urethra	water
glomeruli	renal capsule	urine	

Blood passes into the kidney from the renal _____. The blood is filtered by millions of tangled _____ called _____. Reabsorption of _____ and _____ occurs first in the tubule. _____ are absorbed further down the tubule. The remaining filtrate is called _____. This passes down the _____ to the bladder for storage. [8]

14 Coordination and response

Key objectives

The objectives for this chapter are to revise:
- definitions of the key terms
- nerve impulses and the human nervous system
- the role of the nervous system
- how to identify sensory, relay and motor neurones from diagrams
- simple reflex arcs and how to describe a reflex action
- the synapse as a junction between two neurones
- the structures and functions of the parts of the eye and the pupil reflex
- endocrine glands, their secretions and the functions of adrenaline, insulin, oestrogen and testosterone
- the differences between nervous and hormonal control systems
- the maintenance of a constant internal body temperature

- the role of insulin
- investigations into gravitropism and phototropism in shoots and roots

- the structure of a synapse and how it works
- how to explain the pupil reflex and accommodation
- the distribution and function of rods and cones
- the role of the pancreas in secreting glucagon
- the role of adrenaline
- the concepts of homeostasis and negative feedback, with reference to a set point
- the control of glucose content of the blood
- the symptoms and treatment of type 1 diabetes
- structures in the skin, and its role in the maintenance of a constant internal body temperature
- how to explain gravitropism and phototropism in a shoot, and the role of auxin

Key terms

REVISED ☐

Term	Definition
Gravitropism	A response in which parts of a plant grow towards or away from gravity
Homeostasis	The maintenance of a constant internal environment
Hormone	A chemical substance, produced by a gland and carried by the blood, which alters the activity of one or more specific target organs
Phototropism	A response in which parts of a plant grow towards or away from the direction of the light source
Reflex action	A means of automatically and rapidly integrating and coordinating stimuli with the responses of effectors (muscles and glands)
Sense organ	A group of receptor cells responding to specific stimuli, such as light, sound, touch, temperature and chemicals
Synapse	A junction between two neurones
Set point	The physiological value around which the normal range fluctuates

Nervous control in humans

REVISED ☐

The human nervous system is responsible for the coordination and regulation of body functions. It is made up of two parts:

- the central nervous system (CNS) – brain and spinal cord, the role of which is coordination
- the peripheral nervous system (PNS) – nerves, which connect all parts of the body to the central nervous system.

Sense organs are linked to the peripheral nervous system. They contain groups of receptor cells. When exposed to a stimulus, they generate an **electrical signal** that passes along peripheral nerves to the central nervous system, triggering a response.

Nerve cells (neurones)

Nerves contain nerve cells called **neurones**. Peripheral nerves contain sensory and motor neurones:

- **Sensory neurones** transmit nerve impulses from sense organs to the central nervous system.
- **Motor (effector) neurones** transmit nerve impulses from the central nervous system to effectors (muscles or glands).
- **Relay (connector) neurones** (also called multipolar neurones) are found inside the central nervous system. They make connections with other neurones.

Figure 14.1 shows the structures of neurones. You need to be able to recognise them from their features. Sensory and motor neurones are covered with a myelin sheath, which insulates the neurone to make transmission of the impulse more efficient.

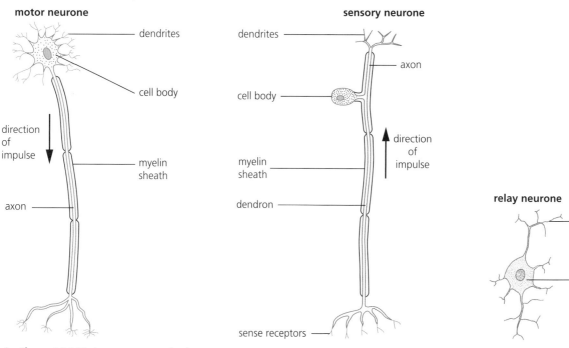

▲ Figure 14.1 Motor, sensory and relay neurones

The cytoplasm (mainly axon or dendron) is elongated in sensory and motor neurones to transmit the impulse for long distances.

Table 14.1 compares the structures of sensory, relay and motor neurones.

▼ Table 14.1 The structure of sensory, relay and motor neurones

Structure	Sensory neurone	Relay neurone	Motor neurone
Cell body	Near the end of the neurone in a ganglion (swelling) just outside the spinal cord	In the centre of the neurone inside the spinal cord	At the start of neurone inside the grey matter of the spinal cord
Dendrites	Present at the end of the neurone	Present at both ends of the neurone	Attached to the cell body

Structure	Sensory neurone	Relay neurone	Motor neurone
Axon (part of neurone taking impulses away from cell body)	Very short	None	Very long
Dendron	Very long	None	None

Sample question

Figure 14.2 shows a type of neurone. Name this type of neurone and state a reason for your choice. [2]

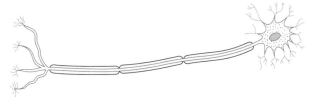

▲ **Figure 14.2**

Student's answer

> Name: motor neurone ✓
>
> Reason: it has a cell body ✗

Teacher's comments

The reason is not enough to distinguish it from other neurones – all neurones have cell bodies. If the answer had been extended to state that the cell body was at the end of the cell, it would have been awarded a mark.

Synapses

Neurones do not connect directly with each other; the junction between two neurones is called a **synapse**. The impulse is 'transmitted' across the synapse by means of a chemical.

At a synapse, a branch at the end of one fibre is in close contact with the cell body or dendrite of another neurone (Figure 14.3).

When an impulse arrives at the synapse, **vesicles** in the cytoplasm release a tiny amount of the neurotransmitter substance. It rapidly diffuses across the synaptic gap (also known as the **synaptic cleft**) and binds with **neurotransmitter receptor proteins** in the membrane of the neurone on the other side of the synapse. This then stimulates an impulse in the neurone.

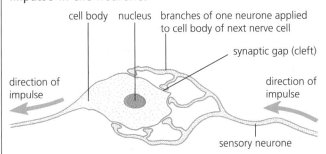

▲ **Figure 14.3 Synapses between neurones**

Synapses control the direction of impulses because neurotransmitter substances are synthesised on only one side of the synapse, while receptor molecules are present only on the other side. They slow down the speed of nerve impulses slightly because of the time taken for the chemical to diffuse across the **synaptic gap**.

The reflex arc

A **reflex arc** describes the pathway of an electrical impulse in response to a stimulus. Figure 14.4 shows a typical reflex arc. The stimulus is a drawing pin sticking in the finger. The response is the withdrawal of the arm caused by contraction of the biceps. Relay neurones are found in the spinal cord, connecting sensory neurones to motor neurones.

The sequence of events is as follows:

Stimulus (sharp pin in finger)
↓
Receptor (pain receptor in skin)
↓
Coordinator (spinal cord)
↓
Effector (biceps muscle)
↓
Response (biceps muscle contracts, hand is withdrawn from pin)

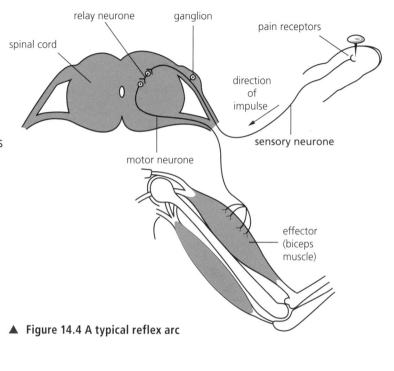

▲ **Figure 14.4 A typical reflex arc**

Sense organs

Table 14.2 gives examples of sense organs and their stimuli.

▼ **Table 14.2 Sense organs and their stimuli**

Sense organ	Stimulus
Ear	Sound, body movement (balance)
Eye	Light
Nose	Chemicals (smell)
Tongue	Chemicals (taste)
Skin	Temperature, pressure, touch, pain

The eye

You need to be able to label parts of the eye on diagrams. Figure 14.5 shows the front view of the left eye and Figure 14.6 shows a section through the eye.

The eyebrow stops sweat running down into the eye. Eyelashes help to stop dust blowing onto the eye. Eyelids can close automatically (blinking is a reflex) to prevent dust and other particles getting onto the surface of the cornea. Blinking also helps to keep the surface moist by moving liquid secretions (tears) over the exposed surface. Tears also contain enzymes that have an antibacterial function.

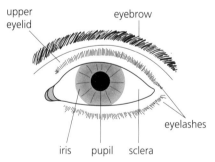

▲ **Figure 14.5 Front view of the left eye**

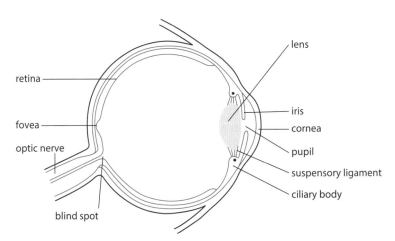

▲ **Figure 14.6 Section through the eye**

Note that the positions of the fovea, ciliary body and suspensory ligament are only needed for the extended syllabus.

Table 14.3 gives the functions of parts of the eye needed for the core paper.

▼ **Table 14.3 Functions of parts of the eye**

Part	Function
Cornea	A transparent layer at the front of the eye that refracts the light entering the eye to help focus it
Iris	A coloured ring of circular and radial muscle that controls how much light enters the pupil
Lens	A transparent, convex, flexible, jelly-like structure that focuses light onto the retina
Retina	A light-sensitive layer containing light receptors, some of which are sensitive to light of different colours
Optic nerve	Carries electrical impulses from the retina to the brain

Pupil reflex

This reflex changes the size of the pupil to control the amount of light entering the eye. In bright light, pupil diameter is reduced, as too much light falling on the retina could damage it. In dim light, pupil diameter is increased to allow as much light as possible to enter the eye.

You need to be able to explain the pupil reflex in terms of light intensity and the antagonistic action of the circular and radial muscles in the iris. **Antagonistic muscles** are those that work in pairs and oppose each other in their actions.

Figure 14.7 shows the effect of light intensity on the iris and pupil.

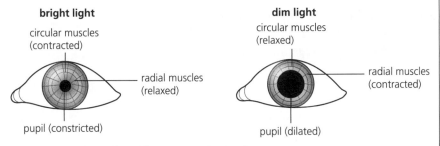

▲ **Figure 14.7 The effect of light intensity on the eye**

The amount of light entering the eye is controlled by altering the size of the pupil. The retina detects the brightness of light entering the eye.

An impulse passes to the brain along sensory neurones and travels back to the muscles of the iris along motor neurones, triggering a response. The change in the size of the pupil is caused by contraction of the radial or circular muscles.

- **High light intensity** causes a contraction in a ring of **circular muscle** in the iris, while radial muscles relax. This reduces the size of the pupil and reduces the intensity of light entering the eye. High-intensity light can damage the retina, so this reaction has a protective function.

- **Low light intensity** causes the circular muscle of the iris to relax and **radial muscle** fibres to contract. This makes the pupil enlarge and allows more light to enter.

Accommodation (focusing)

The amount of focusing needed by the lens depends on the distance of the object being viewed – light from near objects requires a more convex lens than light from a distant object. The shape of the lens needed to accommodate the image is controlled by the ciliary body – this contains a ring of muscle around the lens.

Do not confuse ciliary muscles and circular muscles. Ciliary muscles affect the shape of the lens; circular muscles affect the size of the pupil.

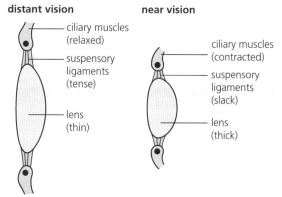

▲ **Figure 14.8 Focusing on distant and near objects**

Distant objects
The ciliary muscles relax, giving them a larger diameter. This pulls on the suspensory ligaments, which, in turn, pull on the lens. This makes the lens thinner (less convex). As the ciliary muscles are relaxed, there is no strain on the eye (Figure 14.8, left-hand side).

Near objects
The ciliary muscles contract, giving them a smaller diameter. This removes the tension on the suspensory ligaments which, in turn, stop pulling on the lens. The lens becomes thicker (more convex) (Figure 14.8, right-hand side). A thicker lens refracts the light more than a thin lens. As the ciliary muscles are contracted, there is strain on the eye, which can cause a headache if a near object (book, microscope, computer screen etc.) is viewed for too long.

Retina
Rods and cones are light-sensitive cells in the retina. When stimulated, they generate electrical impulses, which pass to the brain along the optic nerve. Cones are most concentrated in the fovea. This is the point on the retina where the light is usually focused. Figure 14.6 shows the position of the fovea.

Table 14.4 shows the main differences between rods and cones.

▼ **Table 14.4 Comparison of rods and cones**

Cell	Function	Distribution	Comments
Rods	Sensitive to low light intensity; they detect shades of grey	Found throughout the retina, but none in the centre of the fovea or in the blind spot	These cells provide us with night vision, when we can recognise shapes but not colours
Cones	Sensitive only to high light intensity; they detect colour (but do not operate in poor light)	Concentrated in the fovea	There are three types: cells that are sensitive to red light, green light and blue light

Hormones

Hormones are defined at the start of this chapter. You need to be able to identify four endocrine glands and the hormones that they secrete (Figure 14.9):

- **Adrenaline** is secreted into the blood by the adrenal glands, which are found just above each kidney.
- **Insulin** is secreted by the pancreas. It reduces levels of blood sugar when it gets above normal by instructing the liver to store it, thus removing it from the blood.
- **Oestrogen** is secreted by the ovaries. It prepares the uterus for the implantation of an embryo by making its lining thicker and increasing its blood supply.
- **Testosterone** is secreted by the testes. It plays a part in the development of male secondary sexual characteristics.

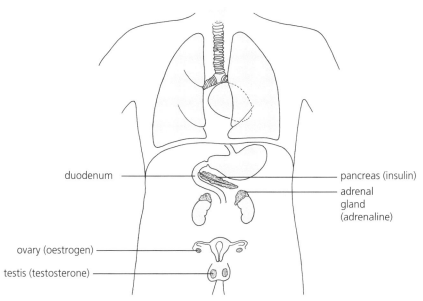

duodenum

pancreas (insulin)

adrenal gland (adrenaline)

ovary (oestrogen)

testis (testosterone)

▲ **Figure 14.9 Position of endocrine glands in the body**

The pancreas also secretes the hormone glucagon. Glucagon changes glycogen to glucose when blood glucose levels fall below normal.

Adrenaline

Adrenaline is a hormone that causes heart rate and breathing rate to increase, and also increases the diameter of the pupils. The muscles are prepared for action in a situation such as a fight or struggle, when needing to run away from danger, a sudden shock or a stressful situation – for example, taking an exam or giving a public performance (hence the expression 'fight or flight').

Control of metabolic activity by adrenaline

The term **metabolism** describes all the chemical changes that take place in the body.

Adrenaline causes the heart rate to increase, so that muscles are supplied with blood containing glucose and oxygen more quickly. This prepares them for action. It also stimulates the liver to convert glycogen to glucose, increasing the blood glucose concentration. Adrenaline also reduces the blood supply to the skin and digestive organs, so blood is diverted to vital organs.

Comparing nervous and hormonal control systems

Table 14.5 shows the main differences between the nervous and hormonal control systems.

▼ **Table 14.5 Comparison of nervous and hormonal control systems**

Feature	Nervous system	Hormonal (endocrine) system
Form of transmission	Electrical impulses	Chemical (hormones)
Transmission pathway	Nerves	Blood vessels
Speed of transmission	Fast	Slow
Duration of effect	Short term	Long term
Response	Localised	Widespread (although there may be a specific target organ)

Sample question   REVISED ☐

Figure 14.10 shows external views of the eye before and after a person has experienced a sudden shock.

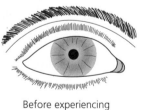
Before experiencing sudden shock

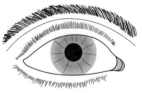
After experiencing sudden shock

▲ **Figure 14.10**

a State the hormone responsible for causing this change.   [1]
b i   Describe the change to the central part of the eye, and how it has happened.   [2]
   ii Suggest how these changes could affect the person's vision.   [2]

Student's answer

a   Adrenaline ✓
b   i   The pupil has got bigger ✓ because the circular muscles of the iris have contracted ✗.
    ii The person's retina could be damaged ✓ because too much light could reach it ✓.

Teacher's comments

The answer to part a is correct. The question illustrates a 'fight or flight' situation.

In part b(i) the student correctly identifies the pupil as the central part of the eye and observes that it has become bigger as a result of the shock, but has confused the circular muscles with the radial muscles of the iris. When the circular muscles contract, the pupil gets smaller, not larger.

The response to part b(ii) is excellent, gaining both marks.

Homeostasis

REVISED ☐

Homeostasis is the process of maintaining a constant internal environment, which is vital for an organism to stay healthy. Fluctuations in temperature, water levels and nutrient concentrations, for example, could lead to death.

Insulin plays an important part in homeostasis because it decreases blood glucose concentrations.

Homeostasis and negative feedback

Homeostasis is the control of internal conditions within set limits. There is a **set point** around which the normal range fluctuates. A change from normal, for instance an increase in temperature, triggers a sensor, which stimulates a response in an effector. However, the response – in this case an increase in sweating and vasodilation of arterioles – would eventually result in temperature levels dropping below normal. As temperature levels drop, the sensor detects the drop and instructs an effector (the skin) to reduce sweating and reduce vasodilation of arterioles. This is negative feedback – the change is fed back to the effector.

Controlling the levels of blood glucose

The liver is a homeostatic organ – it controls the levels of a number of materials in the blood, including glucose. Two hormones – insulin and glucagon – control blood glucose levels. Both hormones are secreted by the pancreas and are transported to the liver in the bloodstream. Excess glucose is stored in the liver and muscles as the polysaccharide glycogen (animal starch). When glucose levels drop below normal, glycogen is broken down to glucose, which is released into the bloodstream.

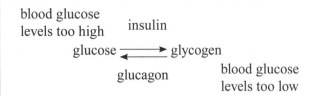

Be accurate with the spellings of *glycogen* and *glucagon*, and make sure you do not get these terms confused.

Type 1 diabetes

There are two types of diabetes. Type 1 is the less common form, which results from a failure of cells of the pancreas to produce enough insulin. The outcome is that the patient's blood is deficient in insulin, and they are unable to regulate the level of glucose in their blood.

Symptoms include feeling tired, feeling very thirsty, frequent urination and weight loss. Weight loss is experienced because the body starts to break down muscle and fat. Blood glucose may rise to such a high level that it is excreted in the urine, or may fall so low that the brain cells cannot work properly and the person goes into a coma.

Diabetics need a carefully regulated diet to keep blood sugar within reasonable limits and to take regular exercise. People with type 1 diabetes also need to monitor their blood sugar levels with frequent blood tests, and may need regular injections of insulin to control them, and thus lead a normal life.

Skin structure

You need to be able to name and identify a number of structures in the skin: hairs, hair erector muscles, sweat glands, receptors (e.g. pressure, temperature, touch, pain), sensory neurones, blood vessels and fatty tissue. These are shown in Figure 14.11.

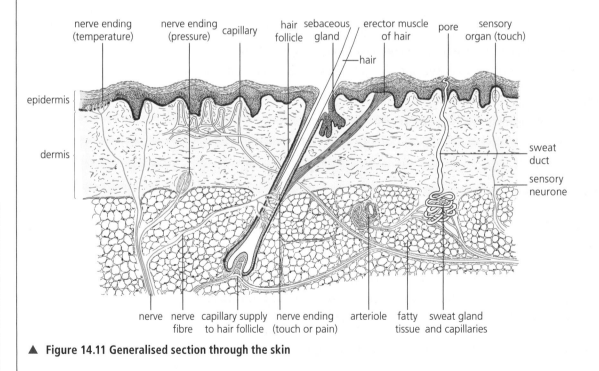

▲ **Figure 14.11 Generalised section through the skin**

The skin and temperature control

Temperature regulation is a homeostatic function. Mammals and birds are warm-blooded – they maintain a constant body temperature despite external environmental changes. Humans maintain a body temperature of 37°C – we have mechanisms to lose heat when we get too hot and ways of retaining heat when we get too cold. Figure 14.12 summarises two ways of regulating body temperature.

Sweating

Sweat is a liquid made up of water, mineral ions and some urea. Sweat glands in the skin secrete sweat through pores onto the skin surface. As the water in the sweat evaporates, it removes heat from the skin, cooling it down. When we are too hot, the volume of sweat produced increases. If we get too cold, the amount of sweat produced is reduced, so less heat is lost through evaporation.

Note that it is just the water from the sweat that evaporates – solutes in the sweat, such as urea and salts, remain on the skin surface.

Insulation

The skin has a layer of fatty tissue that has insulating properties – it reduces heat loss from the skin surface.

Shivering

Uncontrollable bursts of rapid muscular contraction in the limbs release heat as a result of respiration in the muscles. However, its effectiveness in temperature control is questionable.

Role of the brain

The brain plays a direct role in detecting any changes from normal by monitoring the temperature of the blood. A region called the **hypothalamus** contains a thermoregulatory centre, in which receptors detect temperature changes in the blood and coordinate a response to them.

Temperature receptors are also present in the skin. They send information to the brain about temperature changes.

Vasodilation and vasoconstriction

Heat is transported around the body in the bloodstream (see Chapter 9). When blood passes through blood vessels near the skin surface, heat is lost by radiation.

Arterioles (small arteries) have muscle in their walls. When we are too hot, these muscles relax, creating a wide lumen through which lots of blood can pass to the skin surface capillaries (the skin of a hot person may look red). This is called **vasodilation**. More heat is radiated, so we cool down.

When we are too cold, the muscles contract, creating a narrow lumen through which little blood can pass (the skin of a cold person may look very pale). This is called **vasoconstriction**. Less heat is radiated to conserve heat.

Remember that the processes of vasodilation and vasoconstriction only happen in arterioles – they do not take place in capillaries or veins. When writing about the processes, make sure you refer to arterioles.

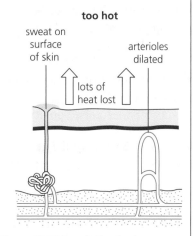

▲ **Figure 14.12 Regulating body temperature**

Tropic responses

Investigations can be carried out to study the effects of **gravitropism** and **phototropism** in shoots and roots of plants. Young seedlings and germinating seeds make good subjects for investigations, because they are cheap, easy to prepare and show results quickly.

- Shoots grow towards light (**positive phototropism**) and away from gravity (**negative gravitropism**).
- Roots grow away from light (**negative phototropism**) and towards gravity (**positive gravitropism**).

Investigating phototropism

One way of investigating the relationship between a plant shoot and light is to select two potted seedlings of similar size and water them both. One is placed in a light-proof box with a window cut in one side, to provide one-sided light (plant A). The other is treated the same way, but set up on a turntable called a clinostat, which slowly turns so that all sides of the shoot receive light (control plant B). The shoot of plant A grows towards the light source, but the shoot of plant B grows vertically. The results suggest that shoot A has responded to one-sided light by growing towards it – a response called positive phototropism.

Investigating gravitropism

One way of investigating the relationship between plant roots and gravity is to select a number of seedlings with straight radicles (young roots) and set them up as shown in Figure 14.13.

The radicles in the clinostat (the control) continue to grow horizontally, but those in the fixed jar grow vertically downwards. It can be concluded that the stationary radicles have responded to gravity by growing towards it – a response called positive gravitropism.

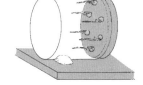

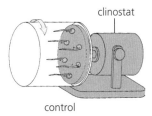
clinostat

control

▲ **Figure 14.13 Investigating gravitropism in roots**

Control of plant growth by auxins

Auxins are plant growth substances. They are sometimes referred to as hormones, but this is not accurate because they are not secreted by glands and are not transported in blood. They are produced by the shoot and root tips of actively growing plants. An accumulation of auxin in a shoot stimulates cell growth by the absorption of water. However, auxins have the opposite effect in roots – when they build up, they slow down cell growth.

You need to be able to describe the role of auxins in phototropism and gravitropism.

Light

When a shoot is exposed to light from one side, auxins that have been produced by the tip move towards the shaded side of the shoot (or the auxins are destroyed on the light side, causing an unequal distribution). Cells on the shaded side are stimulated to absorb more water than those on the light side, so the unequal growth causes the stem to bend towards the light (positive phototropism).

If a root is exposed to light in the absence of gravity, auxins that have been produced by the tip move towards the shaded side of the root.

Cells on the shaded side are stimulated to absorb less water than those on the light side, so the unequal elongation causes the root to bend away from the light (negative phototropism).

Gravity

Shoots and roots also respond to gravity. If a shoot is placed horizontally in the absence of light, auxins accumulate on the lower side of the shoot, owing to gravity. This makes the cells on the lower side elongate more quickly than those on the upper side, so the shoot bends upwards (negative gravitropism).

If a root is placed horizontally in the absence of light, auxins accumulate on the lower side of the root, owing to gravity. However, this makes the cells on the lower side elongate more slowly than those on the upper side, so the root bends downwards (positive gravitropism).

Shoots and roots that have their tips removed will not respond to light or gravity because the part that produces auxins has been cut off. Shoots that have their tips covered with opaque material grow straight upwards when exposed to one-sided light because the auxin distribution is not influenced by the light.

Exam-style questions

1. a Describe the process of transmitting a nerve impulse between two neurones. [6]
   b Explain:
      i   how a synapse ensures that nerve impulses travel in one direction only [2]
      ii why nerve impulses are slower when synapses are involved [2]

2. Figure 14.14 shows a nerve cell.

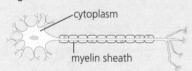

▲ **Figure 14.14**

   a i   Name the type of nerve cell shown in Figure 14.14. [1]
      ii State two features that distinguish it from other types of nerve cell. [2]
      iii Where in the nervous system is this cell located? [1]
   b Nerve cells are specialised cells. Suggest how the cytoplasm and myelin sheath of the nerve cell, labelled in Figure 14.14, enable the nerve cell to function successfully. [4]
   c Reflexes involve a response to a stimulus.
      i   Copy and complete the flowchart below by putting the following terms in the boxes to show the correct sequence in a reflex. [2]

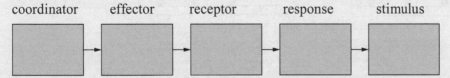

      ii For the pupil reflex, identify each of the parts of the sequence by copying and completing the table below. The first row has been done for you. [4]

Part of sequence	Part in pupil reflex
Coordinator	Brain
Effector	
Receptor	
Response	
Stimulus	

3 Describe and explain how the eye changes its focus from a distant object to a near object.   [5]

4 Copy and complete the following table to show the differences between nervous and hormonal control in the human body.   [4]

Feature	Nervous control	Hormonal control
Speed	Extremely rapid	
Pathway	Neurones	
Nature of 'impulse'		Chemical
Origin		Endocrine gland

5 Copy and complete the paragraph using some of the words given below. Each word may be used once, more than once, or not at all.

excretion   glucose   glycogen   glucagon   insulin   liver

oestrogen   pancreas   secretion   starch   stomach   sucrose

The bloodstream transports a sugar called _____. The blood sugar level has to be kept constant

in the body. If this level falls below normal, a hormone called _____ is released into the

blood by an endocrine organ called the _____. The release of a substance from a gland is

called _____.

Glucagon promotes the breakdown of _____ to increase the blood sugar level. If the blood

sugar level gets too high, the endocrine organ secretes another hormone called _____ into

the blood. This hormone promotes the removal of sugar from the blood and its conversion to

glycogen in the _____.   [7]

6 In Figure 14.16, the left-hand side shows an experiment in which the coleoptiles (shoots) of similar seedlings have been treated in different ways, and the right-hand side shows the result for shoot D 24 hours later.

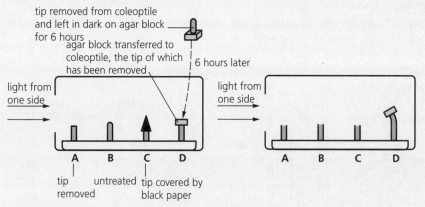

tip removed from coleoptile and left in dark on agar block for 6 hours

agar block transferred to coleoptile, the tip of which has been removed

6 hours later

light from one side

light from one side

A   B   C   D

A   B   C   D

tip removed   untreated   tip covered by black paper

▲ Figure 14.16

a i Name the response shown by shoot D.   [2]
   ii Explain what has caused this response.   [3]
b Copy and complete the right-hand side of Figure 14.16 to show the likely results for shoots A, B and C.   [3]

15 Drugs

Key objectives

The objectives for this chapter are to revise:
- definitions of the key terms
- the use of antibiotics to treat bacterial infections
- that some bacteria are resistant to antibiotics
- that antibiotics do not affect viruses
- how to minimise the development of antibiotic-resistant bacteria

Key term

REVISED

Term	Definition
Drug	Any substance taken into the body that modifies or affects chemical reactions in the body

Drugs

REVISED

Drugs may be used to treat disease, reduce the sensation of pain or help calm us down. In addition, they may change our mood by affecting the brain.

Antibiotics

Antibiotics destroy bacteria without harming the tissues of the patient. This makes them ideal for treating bacterial infections. Most of the antibiotics we use come from bacteria or fungi that live in the soil. One of the best-known antibiotics is **penicillin**, which is produced by the mould fungus *Penicillium*.

Antibiotics attack bacteria in a variety of ways. Some of them prevent the bacteria from reproducing or even cause them to burst open; some interfere with protein synthesis to stop bacterial growth.

Not all bacteria are killed by antibiotics. Some bacteria can mutate into forms that are resistant to these drugs, reducing the effectiveness of antibiotics.

Sample question

REVISED

Figure 15.1 shows the effect of the presence of a mould fungus on bacterial colonies growing on nutrient agar in a Petri dish.

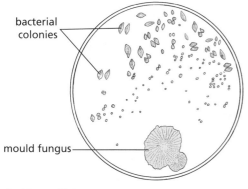

bacterial colonies

mould fungus

▲ Figure 15.1

A student suggested that the mould fungus could be secreting a chemical that diffused into the agar jelly and killed bacterial colonies near it.

a i Suggest what the chemical could be. [1]

   ii What evidence suggests that the concentration of the chemical became lower as it diffused further from the fungus? [1]

b Suggest an alternative reason for why there are no bacterial colonies growing very near the mould fungus. [1]

Student's answer

a i   Antibiotic ✓
   ii   There are bacteria growing near the edge of the plate. ✗
b   The fungus had used up all the nutrients in the agar jelly close to it. ✓

Teacher's comments

The student answered part a(i) correctly. Since it was a 'suggest' question, a named antibiotic such as penicillin would also have been acceptable. The answer to part a(ii) did not answer the question. A mark-worthy answer would be that the bacterial colonies were larger at the edge of the plate.

The answer to part b is good. The stem of the question referred to nutrient agar. The student had read this carefully and used the information to form a sound biological explanation.

Antibiotics and viral diseases

Antibiotics are not effective against viral diseases. This is because antibiotics work by disrupting structures in bacteria such as cell walls and membranes, or processes associated with protein synthesis and replication of DNA. Viruses have totally different characteristics to bacteria, so antibiotics do not affect them.

Development of resistant bacteria

It is really important that the development of resistant bacteria is minimised, because otherwise antibiotics will become ineffective in the treatment of common bacterial infections. MRSA (methicillin-resistant *Staphylococcus aureus*) is one bacterium that is resistant to antibiotics. It is sometimes called a 'superbug'.

To reduce the development of resistant bacteria:

- antibiotics should be used only when essential, otherwise there could be a build-up of a resistant strain of bacteria; the drug resistance can be passed from harmless bacteria to pathogens
- a course of antibiotics should always be completed and not used in a diluted form; otherwise, bacteria that have been exposed to the antibiotic but not killed may mutate into resistant forms

Revision activity

Draw an annotated diagram to show how resistant bacteria such as MRSA can develop.

Exam-style questions

1 a Define the term *drug*. [2]

   b Explain why antibiotics are ineffective against viral diseases. [2]

   c Describe how bacteria can become resistant to antibiotics. [3]

16 Reproduction

Key objectives

The objectives for this chapter are to revise:
- definitions of the key terms
- examples of asexual reproduction
- parts of insect-pollinated and wind-pollinated flowers – their functions and adaptations
- pollination and fertilisation
- differences between the pollen grains and flowers of insect-pollinated and wind-pollinated plants
- investigations into the environmental conditions that affect seed germination
- parts of the male and female human reproductive systems and their functions
- adaptive features of sperm and egg cells
- fertilisation in humans
- early development, growth and development of the fetus
- roles of oestrogen and testosterone in puberty
- the menstrual cycle
- HIV – its transmission and control of its spread

- advantages and disadvantages of asexual and sexual reproduction
- the terms *haploid* and *diploid*
- *self-pollination* and *cross-pollination*, and their potential effects on a population
- the process leading up to fertilisation in a flower
- functions of the placenta and umbilical cord
- the production and roles of hormones in the menstrual cycle and in pregnancy

Key terms

REVISED

Term	Definition
Asexual reproduction	The process resulting in the production of genetically identical offspring from one parent
Fertilisation	The fusion of the nuclei from a male gamete (sperm) and a female gamete (egg cell)
Pollination	The transfer of pollen grains from the anther to a stigma
Sexual reproduction	A process involving the fusion of gametes (sex cells) to form a zygote and the production of offspring that are genetically different from each other
Sexually transmitted infection (STI)	An infection that is transmitted through sexual contact
Cross-pollination	The transfer of pollen grains from the anther of a flower to the stigma of a flower on a different plant of the same species
Self-pollination	The transfer of pollen grains from the anther of a flower to the stigma of the same flower, or a different flower on the same plant

Asexual reproduction

REVISED

Examples of organisms that demonstrate **asexual reproduction** include bacteria, fungi and potatoes. These are described in Figure 16.1.

Bacteria	Fungi	Potatoes
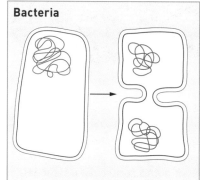	spores sporangium	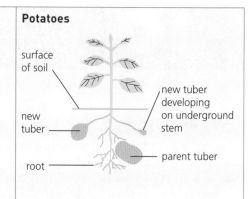
Bacteria reproduce asexually by binary fission. Inside an individual bacterium, the DNA replicates. Then the cell divides into two, with each daughter cell containing a copy of the parental DNA. Once the daughter cells have grown, they can also reproduce	Fungi can reproduce asexually by producing spores, which may be formed inside a structure called a sporangium. When ripe, the sporangium bursts open allowing the spores to be dispersed. In suitable conditions the spores germinate and grow to form new individuals	Potatoes are stem tubers. The parent plant photosynthesises and stores the food produced in underground stems, which swell to form tubers. Each tuber contains stored starch, and there are buds in depressions in the surface known as eyes. In suitable conditions the buds use the stored food to form shoots, from which roots also develop. Each tuber can form a new plant

▲ Figure 16.1 Examples of asexual reproduction

▼ Table 16.1 The advantages and disadvantages of asexual reproduction

Advantages	Disadvantages
The process is quick	There is little variation created, so adaptation to a changing environment (evolution) is unlikely
Only one parent is needed	
No gametes are needed	If the parent has no resistance to a particular disease, none of the offspring will have resistance
All the good characteristics of the parent are passed on to the offspring	
Where there is no dispersal (e.g. in potato tubers), offspring will grow in the same favourable environment as the parent	Lack of dispersal (e.g. potato tubers) can lead to competition for nutrients, water and light
Plants that reproduce asexually usually store large amounts of food that allow rapid growth when conditions are suitable	

You need to be able to relate the advantages and disadvantages outlined in Table 16.1 to a population in the wild and to crop production. In **agriculture** and **horticulture**, asexual reproduction is exploited to preserve desirable qualities in crops: the good characteristics of the parent are passed on to all the offspring; the bulbs produced can be guaranteed to produce the same shape and colour of flower from one generation to the next. In some cases, such as tissue culture, the young plants grown can be transported much more cheaply than, for example, potato tubers, as the latter are much bulkier. The growth of new plants by asexual reproduction tends to be a quick process.

In the wild, it might be a disadvantage to have no variation in a species. If the climate or other conditions change, or a vegetatively produced plant has no resistance to a particular disease, the whole population could be wiped out.

Sexual reproduction

The definitions of the terms **sexual reproduction** and **fertilisation** are given at the start of this chapter. You need to learn these.

You need to be able to state that the nuclei of gametes (sex cells) are haploid (see Chapter 17) and that the nucleus of the zygote (a fertilised egg cell) is diploid.

▼ **Table 16.2 The advantages and disadvantages of sexual reproduction**

Advantages	Disadvantages
There is variation in the offspring, so adaptation to a changing or new environment is likely to occur, enabling survival of the species	Two parents are usually needed (although not always – some plants can self-pollinate)
New varieties can be created, which may have resistance to disease	Growth of a new plant to maturity from a seed is slow
In plants, seeds are produced, which allows dispersal away from the parent plant, reducing competition	

You need to be able to relate the advantages and disadvantages outlined in Table 16.2 to a population in the wild and to crop production.

Sexual reproduction is exploited in agriculture and horticulture to produce new varieties of animals and plants by cross-breeding. The new varieties can have the combined features of the organisms used to produce them.

Sexual reproduction in plants

You need to be able to describe the structure and functions of parts of an insect-pollinated flower.

Flower structure and functions

Figure 16.2 shows the main parts of a lupin flower that has been cut in half. Other flowers have the same features, but the numbers and relative sizes of the parts vary.

Table 16.3 outlines the main functions of the parts of a flower.

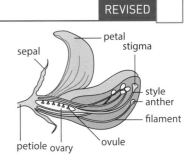

▲ **Figure 16.2 Structure of a lupin flower**

▼ **Table 16.3 Functions of the parts of a flower**

Part	Function
Petal	Often large and coloured to attract insects
Sepal	Protects the flower while in bud
Stamen	The male reproductive part of the flower, made up of the anther and filament
Anther	Contains pollen sacs in which pollen grains are formed – pollen contains male sex cells; note: you need to be able describe an anther
Filament	Supports the anther
Carpel	The female reproductive part of the flower, made up of the stigma, style and ovary
Stigma	A sticky surface that receives pollen during pollination; note: you need to be able to describe a stigma
Style	Links the stigma to the ovary; through which pollen tubes grow
Ovary	Contains ovules
Ovule	Contains a nucleus, which develops into a seed when fertilised

Skills

Drawing specimens

One of the skills you may be tested on in the Practical or Alternative to Practical exam is your ability to draw a specimen or an image in a photograph, such as a flower.

Make sure you follow these points when producing biological drawings. You can lose marks if you do not follow these guidelines:
● Use a sharp HB pencil, with outlines sharp, clear and unbroken – not 'sketchy'.
● Make your drawing as large as will fit into the space available.

● Avoid unnecessary shading or colour because this can often cover important detail.
● Your drawing should show what has been asked for – for example, the whole specimen or one part.
● Show on your drawing any detail that can be observed, with the correct number of parts and details as seen in the specimen.
● The relative proportions of the parts of the specimen should be correctly represented.
● Label your drawing if told to do so.
● Use ruled label lines in pencil; these should touch the object or feature labelled.

Pollination

The definition of **pollination** was given at the start of this chapter. Flowers are usually pollinated by insects or wind. The structural adaptations of a flower depend on the type of pollination the plant uses.

Insect and wind pollination

Figure 16.3 shows the structure of a grass flower that is wind pollinated. Although you do not need to be able to draw this, you do need to be able to compare insect- and wind-pollinated flowers.

Figure 16.4 shows the appearance of wind-borne and insect-borne pollen grains.

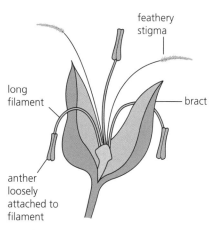

▲ Figure 16.3 Structure of a grass flower

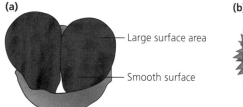

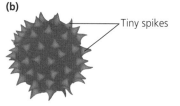

▲ Figure 16.4 Wind-borne (a) and insect-borne (b) pollen grains

Table 16.4 compares the features of insect- and wind-pollinated flowers.

▼ Table 16.4 Comparison of insect- and wind-pollinated flowers

Feature	Insect pollinated	Wind pollinated
Petals	Present – often large, coloured and scented, with guidelines to guide insects into the flower	Absent, or small and inconspicuous
Nectar	Produced by nectaries to attract insects	Absent
Stamen	Present inside the flower	Long filaments, allowing the anthers to hang freely outside the flower so the pollen is exposed to the wind
Stigma	Small surface area, inside the flower	Large and feathery, hanging outside the flower to catch pollen carried by the wind
Pollen	Smaller numbers of grains – grains are often round and sticky, or covered in spikes to attach to the furry bodies of insects	Larger numbers of smooth and light pollen grains, which are easily carried by the wind
Bracts (modified leaves)	Absent	Sometimes present

Note that pollination and seed dispersal are not the same thing. When animals such as insects carry pollen, they aid pollination. When animals carry seeds, they aid seed dispersal.

Sample question

Figure 16.5 shows a flower that is wind pollinated.

a Name structures X and Y. [2]
b Explain how a feature, visible in Figure 16.5, suggests that this
   flower is wind pollinated. [2]

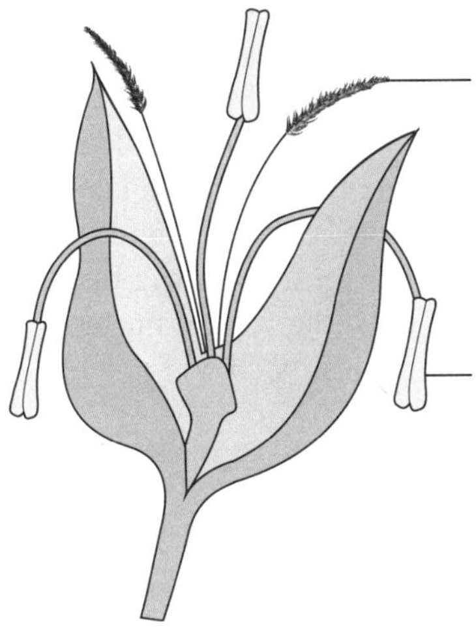

Student's answer

> a X is the stigma ✓; Y is the stamen ✓.
> b Part Y produces a large amount of pollen grains. ✗ These have a large surface
>   area to catch the wind and be carried away to the stigma of another plant. ✗

▲ Figure 16.5

Teacher's comments

In part a, the student has correctly identified part X as the stigma. Part Y is the anther, but the term *stamen* has been awarded a mark because stamen is the unit that includes both the anther and the filament.

In part b, the student has not read the instruction in the question carefully enough: the feature described needs to visible in the diagram. Pollen grains are not shown. Although the answers given are biologically correct, they do not answer the question, so they gain no marks. A correct answer would refer to the stigmas being feathery, or hanging out of the flower to catch pollen blowing past, or to the anthers hanging loose, outside the flower, to allow the pollen to be carried away by the wind.

Correct answer

a X is the stigma, Y is the anther.
b The stigmas are feathery, giving them a larger surface area. They hang out of the flower to catch pollen blowing past.

Self-pollination and cross-pollination

Self-pollination involves the transfer of pollen from the anther to the stigma of the same flower or to another flower of the same plant. A smaller number of pollen grains needs to be produced because there is a greater chance of successful pollination. This increases the chance of fertilisation and seed formation, but reduces variation in the offspring. Self-pollinated plants are less likely to be able to adapt to cope with environmental change.

Cross-pollination involves the transfer of pollen from the anther of a flower to the stigma of a flower on a different plant of the same species. This reduces the chance of fertilisation (wind-pollinated flowers produce large numbers of pollen grains because of the wastage involved), but increases variation and the ability to adapt to environmental change.

Fertilisation

Fertilisation happens when the pollen nucleus fuses with the nucleus of the ovule.

Growth of the pollen tube and the process of fertilisation

Figure 16.6 shows a section through a single carpel. If pollen grains are of the same species as the flower they land on, they may germinate. Germination is triggered by a sugary solution on the stigma and involves the growth of a pollen tube from the pollen grain. The pollen tube contains the male nucleus, which is needed to fertilise the ovule inside the ovary. The pollen tube grows down the style, through the ovary wall and through the micropyle of the ovule. Fertilisation is the fusion of the male nucleus (in the pollen grain) with the female nucleus (in the ovule). If the ovary contains a lot of ovules, each will need to be fertilised by a different pollen nucleus.

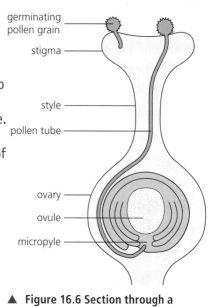

▲ Figure 16.6 Section through a single carpel

Germination

A seed is a living structure. It contains an embryo that will germinate and develop into an adult plant if provided with suitable conditions. These are listed and explained in Table 16.5.

▼ Table 16.5 Environmental conditions required for germination

Environmental condition	Explanation
Water	Water is absorbed through the micropyle until the radicle has forced its way out of the testa
	It is needed to activate enzymes that convert insoluble food stores into soluble foods, which can be used for growth and energy production
Oxygen	Oxygen is needed for respiration to release energy for growth and the chemical changes needed for mobilisation of food reserves
Suitable temperature	Enzymes work best at an optimum temperature – generally, the higher the temperature (up to 40°C), the faster the rate of germination
	However, some seeds need a period of chilling before they will germinate; low temperatures usually maintain dormancy – if the seed germinated in unsuitable conditions, it would be unlikely to survive

Skills

Investigating the environmental conditions required for germination

You need to describe investigations to study the environmental conditions that affect the germination of seeds. When planning this sort of investigation, remember to change only one variable and keep all the others the same. The experiment shown in Figure 16.7 is one way of investigating the need for water. There should be the same number of

the same variety of peas in each container. The containers should be the same size, and both kept at room temperature in the light for the same amount of time.

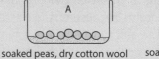
soaked peas, dry cotton wool

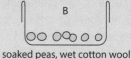
soaked peas, wet cotton wool

▲ Figure 16.7 Experiment to show the need for water in germination

Sexual reproduction in humans

Structure and function of parts of the male reproductive system

Figure 16.8 and Table 16.6 show the structure and function of parts of the male reproductive system.

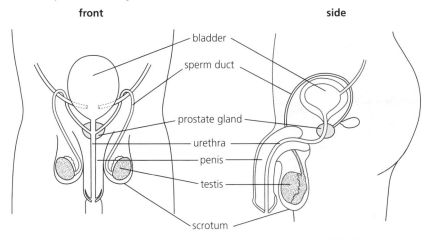

▲ **Figure 16.8 The male reproductive system (also showing the bladder)**

▼ **Table 16.6 Functions of parts of the male reproductive system**

Part	Function
Penis	Can become firm so that it can be inserted into the vagina of the female during sexual intercourse to transfer sperm
Prostate gland	Adds fluid and nutrients to sperm to form semen
Scrotum	A sac that holds the testes outside the body, keeping them cooler than body temperature
Sperm duct	Muscular tube that links the testis to the urethra to allow the passage of semen containing sperm
Testis	Male gonads that produce sperm
Urethra	Passes semen containing sperm to the penis; also carries urine from the bladder at different times

Structure and function of parts of the female reproductive system

Figure 16.9 and Table 16.7 show the structure and function of parts of the female reproductive system.

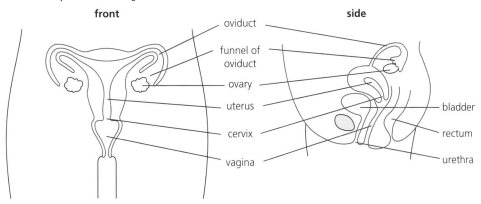

▲ **Figure 16.9 The female reproductive system (also showing the bladder and rectum)**

▼ Table 16.7 Functions of parts of the female reproductive system

Part	Function
Cervix	A ring of muscle that separates the vagina from the uterus
Ovary	Contains follicles in which ova (eggs) are produced
Oviduct	Carries an ovum to the uterus, with propulsion provided by tiny cilia in the wall; also the site of fertilisation
Uterus	Where the fetus develops
Vagina	Receives the male penis during sexual intercourse; sperm are deposited here

Revision activity

Make a set of flashcards for the parts of the male and female human reproductive systems. Include the function for each part. When using the cards for revision, check that you can identify whether the part is male or female, what its function is, and that you know the correct spelling. Note that some parts have similar spellings – for example, ureter and urethra. It is really important to write these correctly and know their functions.

Adaptive features and comparison of sperm and egg cells

Diagrams of sperm and egg cells can be found in Figure 2.5 on page 14. These are compared in Table 16.8.

▼ Table 16.8 Comparison of sperm and egg cells

	Sperm	Egg
Size	Much smaller than an egg cell	Much larger than a sperm cell
Adaptive features	Sperm cells have a flagellum (tail) to swim with The tip of the sperm (called an acrosome) produces enzymes to digest the cells around an egg and the egg membrane The mid-piece (the section of the sperm behind the nucleus) contains numerous mitochondria that provide energy for the movement of the flagellum	Egg cells have a relatively large amount of cytoplasm in which there are energy stores (yolk droplets containing fat) They have a jelly coat that changes after fertilisation
Motility	Flagellum allows the cell to swim to the egg – from the vagina to the oviduct	Not motile
Numbers	Millions present in a single ejaculation	One egg released each month

Fertilisation

Fertilisation is the fusion of the nuclei from a male gamete (sperm) and a female gamete (egg cell/ovum). It occurs in the oviduct. The fertilised egg cell is called a zygote.

Pregnancy and development

The zygote starts to divide by mitosis to form a ball of cells (a blastula). It continues to move down the oviduct until it reaches the uterus. **Implantation** occurs when the blastula embeds in the lining of the uterus.

The blastula develops into an embryo and some of the cells form a placenta, linking the embryo with the uterus lining. Organs such as the heart develop and, after 8 weeks, the embryo is called a fetus. Growth of the fetus requires a good supply of nutrients and oxygen. This is achieved through the link between the placenta and the mother's blood supply in the uterus lining (Figure 16.10). The **placenta** transfers oxygen to the blood supply of the fetus, which passes through the **umbilical cord**, and removes carbon dioxide and other waste products. The fetus is surrounded by an **amniotic sac**, which is a membrane formed from cells of the embryo that contains the amniotic fluid. It encloses the developing fetus and prevents the entry of bacteria.

Amniotic fluid supports the fetus, protecting it from physical damage. It absorbs excretory materials (urine) released by the fetus.

As the fetus grows in the early stages, it becomes increasingly complex, with systems of the body developing. Towards the end of pregnancy, its size increases substantially.

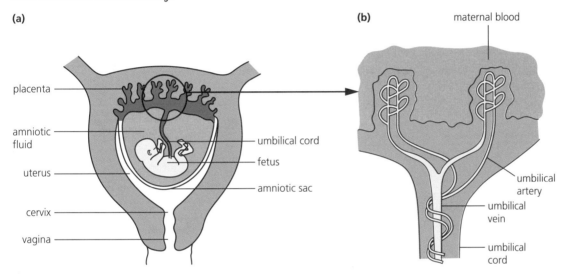

▲ **Figure 16.10 (a) A fetus developing in the uterus. (b) The link between the placenta and the mother's blood supply in the uterus lining**

Functions of the placenta and umbilical cord

The placenta brings the blood supply of the fetus close to that of the mother, but prevents mixing. This is really important because the fetus and mother may have different blood groups – any mixing could result in blood clotting, which could be fatal to both mother and fetus. Blood from the fetus passes through the umbilical cord, in the umbilical artery, to the placenta. Here it comes close to the mother's blood. Oxygen, amino acids, glucose and other nutrients diffuse into the blood of the fetus from the mother's blood. Carbon dioxide, urea and other wastes pass into the mother's blood from the blood of the fetus. Blood returns to the fetus through the umbilical vein, also in the umbilical cord.

The placenta acts as a barrier to toxins and pathogens. However, some toxins, such as drugs (such as aspirin and heroin), along with nicotine and carbon monoxide from smoking, alcohol from drinks and also viruses such as HIV and rubella (German measles), can all pass across the placenta, risking the health of the developing fetus.

Sample question

Figure 16.11 shows a fetus developing in the uterus. Copy and complete the table below by identifying the parts labelled A, B and C, and stating a function of each one. [6]

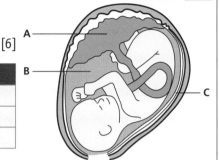

▲ Figure 16.11

Part	Name	Function
A		
B		
C		

Student's answer

Part	Name	Function
A	Placenta ✓	Provides the fetus with blood containing oxygen from the mother ✗
B	Amniotic fluid ✓	Protects the fetus ✗
C	Uterus ✗	Contains the fetus during pregnancy ✗

Correct answer

Part	Name	Function
A	Placenta	Prevents the blood of the mother and fetus from mixing
B	Amniotic fluid	Protects the fetus from physical damage
C	Amniotic sac	Contains the amniotic fluid

Teacher's comments

The description of the function of the placenta is very badly worded – the placenta prevents the blood of the mother and fetus from mixing. Answers containing biologically incorrect information are penalised.

Details about the amniotic fluid are too vague to gain the mark for the function. The correct answer is 'to protect the fetus *from physical damage*'.

Part C is the amniotic sac, which contains the amniotic fluid.

Sexual hormones in humans

Sexual hormones are responsible for the development of secondary sexual characteristics at puberty. Testosterone, secreted by the testes, causes the changes in boys; oestrogen, secreted by the ovaries, causes the changes in girls.

Puberty

Puberty is when the sex organs (ovaries in girls; testes in boys) become mature and start to secrete hormones and make gametes (ova and sperm). Puberty usually happens between the ages of 10 and 14 years, but this varies from person to person.

Table 16.9 shows the secondary sexual characteristics that appear at puberty. A drop in hormone levels can reduce these features, while a high level of hormone can increase them.

▼ Table 16.9 Secondary sexual characteristics that appear at puberty

Male	Female
Voice becomes much lower (breaks)	Breasts grow, nipples enlarge
Hair starts to grow on chest, face, under arms and in pubic area	Hair develops under arms and in pubic area
	Hips become wider
Body becomes more muscular	Uterus and vagina become larger
Penis becomes larger	Ovaries start to release eggs and periods begin (menstruation)
Testes start to produce sperm	

The menstrual cycle

This is a cycle involving changes in the uterus and ovaries, controlled by a number of hormones. Each cycle takes about 28 days:

- At the start of each cycle, menstruation occurs – the lining of the uterus breaks down, and the cells and blood in the lining are shed via the vagina. This is **menstruation**.
- The uterus lining then starts to build up again, developing a mass of blood vessels so that it is ready to receive a fertilised ovum.
- A follicle in one of the ovaries matures into an ovum.
- About half-way through the cycle, the wall of the ovary ruptures and an ovum is released.
- Towards the end of the cycle, the lining of the uterus breaks down again.

Hormones and the menstrual cycle

The sites of production of oestrogen and progesterone in the menstrual cycle and in pregnancy are listed in Table 16.10.

▼ Table 16.10 Sites of production of oestrogen and progesterone

Hormone	Site of production	
	In the menstrual cycle	**In pregnancy**
Oestrogen	Ovaries	Placenta
Progesterone	Corpus luteum (remains of follicle in ovary after ovulation)	Placenta

The events in the menstrual cycle are shown in Figure 16.12.

At the start of the cycle, the lining of the uterus wall has broken down (menstruation). As each follicle in the ovaries develops, the amount of **oestrogen** produced by the ovary increases. The oestrogen acts on the uterus and causes its lining to become thicker and develop more blood vessels. These are changes that help an early embryo to implant. The **pituitary gland** at the base of the brain secretes **follicle-stimulating hormone (FSH)** and **luteinising hormone (LH)**, or **lutropin**, which promote ovulation.

Once the ovum has been released, the follicle that produced it develops into a solid body called the **corpus luteum**. This produces a hormone called **progesterone**, which makes the uterus lining grow thicker and produce more blood vessels. If the ovum is fertilised, the corpus luteum continues to release progesterone and so keeps the uterus in a state suitable for implantation. If the ovum is not fertilised, the corpus luteum stops producing progesterone. The lining of the uterus then breaks down and loses blood, which escapes through the cervix and vagina (menstruation).

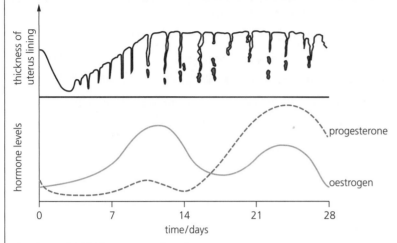

▲ Figure 16.12 The menstrual cycle

Role of hormones in controlling the menstrual cycle and pregnancy

Oestrogen and progesterone control important events in the menstrual cycle. Oestrogen encourages the regrowth of the lining of the uterus wall after a period, and prevents the release of FSH. If FSH is blocked, no further ova are matured. The uterus lining needs to be thick to allow successful implantation of an embryo. Progesterone maintains the thickness of the uterine lining. It also inhibits the secretion of LH, which is responsible for ovulation. If LH is suppressed, ovulation cannot happen, so there are no ova to be fertilised.

Sexually transmitted infections

REVISED

You need to learn the definition of a **sexually transmitted infection (STI)** given at the start of this chapter. These are diseases passed on during unprotected sexual contact. You need to know that **human immunodeficiency virus (HIV)** is a pathogen that causes an STI. HIV may result in acquired immune deficiency syndrome (AIDS). Details are shown in Table 16.11.

▼ Table 16.11 Transmission of HIV and ways to prevent its spread

Methods of transmission	Ways of preventing its spread
Unprotected sexual intercourse with an infected person (this includes homosexuals)	Use of a condom for sexual intercourse
	Abstinence from sexual intercourse
Drug use involving sharing a needle used by an infected person	Use of sterilised needles for drug injections
Transfusions of unscreened blood	Screening of blood used for transfusions
From an infected mother to the fetus	Feeding a baby with bottled milk when the mother has HIV
Feeding a baby with milk from an infected mother	
Use of unsterilised surgical instruments	Use of sterilised surgical instruments

Exam-style questions

1 a Define the term *asexual reproduction*. [2]
   b State the process bacteria use to reproduce. [1]
   c Describe the process of asexual reproduction in fungi. [3]
2 a Name the reproductive structures making up the following parts of a flower:
      i stamen [2]   ii carpel [3]
   b State two ways in which the following parts of a wind-pollinated flower and insect-pollinated flower are different:
      i stamen [2]   ii pollen [2]

3 a Distinguish between self-pollination and cross-pollination. [2]
   b State two advantages of:
      i asexual reproduction [2]
      ii sexual reproduction [2]
4 Figure 16.13 shows a fetus developing in the uterus.
   a Copy the figure and label parts A and B. [2]
   b Outline three functions of the placenta. [3]
   c The blood of the fetus and that of the mother flow close to each other in the placenta, but do not mix. State two advantages to the fetus of having a separate blood system from that of the mother. [2]
5 Figure 16.14 represents part of the male reproductive system, together with parts of the urinary system.
   a Copy or trace the figure and label:
      i the sperm duct (vas deferens) [1]
      ii the urethra [1]
   b What is the difference in function of the urethra between males and females? [2]
   c i The hormone testosterone controls the development of secondary sexual characteristics in males. State two of these characteristics that develop at puberty. [2]
      ii On your drawing, label clearly where this hormone is produced. [1]
      iii Some international athletes, female as well as male, have taken testosterone, illegally, as a drug. Suggest why these athletes might have done this. [2]

6 a HIV can be passed from mother to fetus through the placenta. State two other ways in which the virus can be passed to an uninfected person. [2]
   b Name two other harmful materials that might pass from mother to fetus through the placenta. [2]

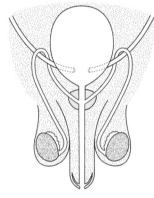

▲ **Figure 16.13**

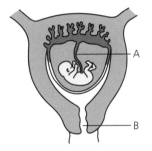

▲ **Figure 16.14**

17 Inheritance

Key objectives

The objectives for this chapter are to revise:
- definitions of the key terms
- what chromosomes are and what they contain
- inheritance of sex in humans
- that a heterozygous individual cannot be pure breeding
- how to interpret pedigree diagrams
- how to use genetic diagrams and Punnett squares to show monohybrid crosses

- the significance of the sequence of bases in a gene
- how DNA controls cell function and how proteins are made
- that all body cells in an organism contain the same genes, but many genes are not expressed
- the number of chromosomes in human haploid and diploid cells

- mitosis and meiosis, including the role of mitosis
- that the exact duplication of chromosomes occurs before mitosis and that, during mitosis, the copies of chromosomes separate
- that meiosis produces variation
- that meiosis is involved in the production of gametes
- how to use a test-cross to identify an unknown genotype
- how to explain codominance, using the inheritance of ABO blood groups
- sex-linked inheritance, using colour blindness as an example
- how to use genetic diagrams to predict the results of monohybrid crosses involving codominance and sex-linkage, and how to calculate phenotypic ratios

Key terms

Term	Definition
Allele	An alternative form of a gene
Dominant	An allele that is expressed if it is present in the genotype
Gene	A length of DNA that codes for a protein
Genotype	The genetic make-up of an organism in terms of the alleles present
Heterozygous	Having two different alleles of a particular gene; therefore, a heterozygous individual will not be pure breeding
Homozygous	Having two identical alleles of a particular gene; two homozygous individuals that breed together will be pure breeding
Inheritance	The transmission of genetic information from generation to generation
Phenotype	The observable features of an organism
Recessive	An allele that is expressed only when no dominant allele of the gene is present in the genotype
Codominance	A situation in which both alleles in a heterozygous organism contribute to the phenotype
Diploid nucleus	A nucleus containing two sets of chromosomes
Haploid nucleus	A nucleus containing a single set of chromosomes
Meiosis	Reduction division in which the chromosome number is halved from diploid to haploid, resulting in genetically different cells
Mitosis	Nuclear division giving rise to genetically identical cells
Sex-linked characteristic	A feature in which the gene responsible is located on a sex chromosome, which makes it more common in one sex than in the other
Stem cell	An unspecialised cell that divides by mitosis to produce daughter cells that can become specialised for specific purposes

Chromosomes, genes and proteins

The definitions of the terms **gene** and **allele** are given at the start of this chapter. You need to learn these.

A chromosome is a thread-like structure of DNA carrying genetic information in the form of genes. Each chromosome is made up of a large number of genes coding for the formation of different proteins that give us our characteristics. Figure 17.1 shows the relationship between a chromosome and the genes it carries.

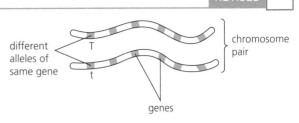

▲ **Figure 17.1 The relationship between a chromosome and the genes it carries**

Diploid nucleus and haploid nucleus

The definitions of **diploid nucleus** and **haploid nucleus** are given at the start of this chapter. In each diploid cell (nearly all body, or somatic, cells) there is a pair of each type of chromosome (Figure 17.1). In a human diploid cell, there are 23 pairs. Sex cells (sperm and ova) are haploid, containing only 23 chromosomes. The 23 chromosomes comprise one from each pair.

The inheritance of sex

Of the 23 pairs of chromosomes present in each human cell, one pair is the sex chromosomes. These determine the sex of the individual. Males have XY, females have XX. So, the presence of a Y chromosome results in male features developing. Figure 17.2 shows how sex is inherited.

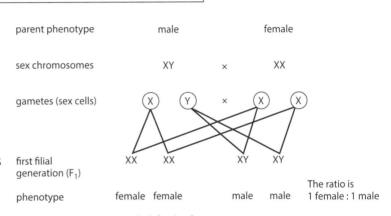

▲ **Figure 17.2 How sex is inherited**

The genetic code

The structure of DNA is described in Chapter 4. In summary:

- Each nucleotide carries one of four bases (A, T, C or G). A string of nucleotides therefore holds a sequence of bases.
- Each group of three bases stands for one amino acid.
- A gene is a sequence of triplets of the four bases, which codes for one protein molecule.
- The sequence of bases forms a code, which instructs the cell to make particular proteins. Proteins are made from amino acids linked together (Chapter 4).
- The types and sequence of amino acids joined together will determine the kind of protein formed and the shape of the protein molecule.
- Insulin is a small protein with only 51 amino acids and therefore 153 (i.e. 3 × 51) bases in the DNA molecule. Most proteins are much larger than this, and most genes contain a thousand or more bases.

The chemical reactions that take place in a cell determine the type of cell it is and what its functions are. These chemical reactions are, in turn, controlled by enzymes. Enzymes are proteins. Therefore, by determining which proteins (particularly enzymes) are produced in a cell, the **genetic code** of DNA also determines the cell's structure and function. In this way, the genes also determine the structure and function of the whole organism. Other proteins coded for in DNA include antibodies, membrane receptors and the receptors for neurotransmitters (see details of synapses in Chapter 14).

The manufacture of proteins in cells

DNA molecules remain in the nucleus, but the proteins that they carry the codes for are needed elsewhere in the cell.

- A molecule called **messenger RNA (mRNA)** is used to transfer the information from the nucleus. mRNA is a copy of a gene.
- An RNA molecule is much smaller than a DNA molecule and is made up of only one strand.
- To pass on the protein code, the double helix of DNA unwinds to expose the chain of bases.
- One strand acts as template. An mRNA molecule is formed along part of this strand, made up of a chain of nucleotides with complementary bases to a section of the DNA strand.
- The mRNA molecule carrying the protein code then moves out of the nucleus into the cytoplasm, where it passes through a ribosome.
- The mRNA molecule instructs the ribosome to put together a chain of amino acids in a specific sequence, thus making a protein. Other mRNA molecules will carry codes for different proteins.

Gene expression

Body cells do not all have the same requirements for proteins. For example, the function of some cells in the stomach is to make the protein pepsin (see 'Chemical digestion' in Chapter 7). Bone marrow cells make the protein haemoglobin, but do not need digestive enzymes. Specialised cells all contain the same genes in their nuclei, but only the genes needed to code for specific proteins are switched on (expressed). This enables the cell to make only the proteins it needs to fulfil its function.

Mitosis

REVISED

The definition of **mitosis** is given at the start of this chapter. Mitosis is a form of cell division used for making new, genetically identical, cells to enable growth or the replacement of old or damaged cells. Asexual reproduction involves mitosis.

Before the process starts, all the chromosomes are duplicated exactly. During mitosis, the copies of the chromosomes separate and form two nuclei with the same number of chromosomes as the parent nucleus cell (the diploid number of chromosomes is maintained). At the end of a mitotic cell division, the number of cells is doubled and the daughter cells produced are genetically identical to the parent.

Although many textbooks show the stages of mitosis, you do not need those details for the Cambridge IGCSE core or extended exam.

Stem cells

Stem cells are unspecialised cells in the body that have retained their power of division by mitosis. The daughter cells produced can become specialised for specific functions. Examples include the basal cells of the skin, which keep dividing to make new skin cells, and cells in the red bone marrow, which divide constantly to produce the whole range of blood cells.

Meiosis

The definition of **meiosis** is given at the start of this chapter. Sex cells (gametes) are formed in the gonads (ovaries and testes) by meiosis. When ova are formed in a woman, all the ova will carry an X chromosome. When sperm are formed in a man, half the sperm will carry an X chromosome and half will carry a Y chromosome (Figure 17.3).

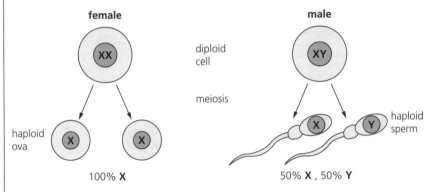

▲ Figure 17.3 **The formation of sex cells by meiosis**

Meiosis is called a reduction division because it involves halving the normal chromosome number – the pairs of chromosomes are separated. The gametes (sex cells) produced are haploid, but they are formed from diploid cells.

At the end of the process, the cells produced are not all identical – meiosis results in genetic variation. Both the maternal and paternal chromosomes contain new combinations of genetic material.

Although many textbooks show the stages of meiosis, you do not need those details for the Cambridge IGCSE core or extended exam.

Sample question

Complete the following passage, using only words from the list below.

diploid   gametes   haploid   meiosis   mitosis   red blood cells

The transfer of inherited characteristics to new cells and new individuals depends on two types of cell division. During _____ the chromosomes are duplicated exactly and _____ cells are produced.

However, during _____ the chromosome sets are first duplicated and then halved, producing haploid cells. These cells will become

_____.   [4]

Student's answer

During **meiotosis ✗**, the chromosomes are duplicated exactly and **identical ✗** cells are produced.

However, during **meiosis ✓**, the chromosome sets are first duplicated and then halved, producing haploid cells. These cells will become **gametes**. **✓**

Revision activity

Construct a table to compare the features of mitosis and meiosis.

Teacher's comments

The first answer is not clear – it mixes up the terms 'mitosis' and 'meiosis'. Sometimes students do this deliberately when they are not sure of the answer, hoping that they will be given the benefit of the doubt – they will not. This student has not followed the instructions in the question for the second answer: the term 'identical' does not appear in the word list. The correct answers are 'mitosis' and 'diploid'.

Monohybrid inheritance

REVISED ☐

You need to learn, and be able to use, definitions of seven genetic terms: **inheritance**, **genotype**, **phenotype**, **homozygous**, **heterozygous**, **dominant** and **recessive**. Definitions of these terms are given at the start of this chapter.

Monohybrid inheritance involves the study of how a single gene is passed on from parents to offspring. It is probably easiest to predict the outcome of a monohybrid cross using a Punnett square (Figure 17.4). However, if you have been taught the traditional way of displaying the cross (as shown in Figure 17.2), there is nothing wrong with using that method.

All the genetic crosses shown will involve examples using pea plants, which can be tall (**T**) or dwarf (**t**) – tall is dominant to dwarf.

Skills

Constructing a Punnett square
- Draw a box that has four compartments.
- Above the top boxes, identify the gametes of one parent, for example the male.
- On the left side of the boxes, identify the gametes of the other parent, for example the female.
- Circle the letters of the gametes to identify them as gametes.
- Complete the Punnett square by writing in the genotype of each of the four offspring produced. Do this by writing in the corresponding gamete

alleles from above the box and from the left of the box (Figure 17.4). Complete all four boxes.

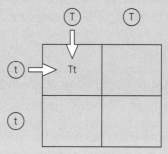

▲ **Figure 17.4 A Punnett square**

Skills

Writing out a genetic cross
- Make sure you state what the symbols represent – for example, **T** = tall, **t** = dwarf.
- If you are given letters for alleles in genetics questions, do not ignore them and make up

your own. This usually results in a number of marks being lost through errors that could easily have been avoided.
- Make sure you label each line in the cross (phenotype, genotype etc.).
- It is a good idea to circle the gametes to show that meiosis has happened.

If two identical homozygous individuals are bred together, the product of the cross will be pure breeding. However, if one parent is pure-breeding tall and the other parent is pure-breeding dwarf, there will be a different outcome, as shown in the following example.

A cross between a pure-breeding tall pea plant and a pure-breeding dwarf pea plant

As tall is dominant to dwarf, and both plants are pure breeding, their genotypes must be **TT** and **tt** (Figure 17.5).

phenotypes of parents	**tall**		**dwarf**

genotypes of parents	**TT**	×	**tt**

gametes (T) (T) × (t) (t)

Punnett square

	(T)	(T)
(t)	**Tt**	**Tt**
(t)	**Tt**	**Tt**

F_1 genotypes — all **Tt**

F_1 phenotypes — all tall

▲ Figure 17.5

A cross between two heterozygous tall pea plants

The genotype of both plants must be **Tt** (Figure 17.6).

phenotypes of parents	**tall**	**tall**

genotypes of parents	**Tt**	×	**Tt**

gametes (T) (t) × (T) (t)

Punnett square

	(T)	(t)
(T)	**TT**	**Tt**
(t)	**Tt**	**tt**

F_1 genotypes — 1 **TT**, 2 **Tt**, 1 **tt**

F_1 phenotypes — **tall tall dwarf**

ratio — 3 tall : 1 dwarf

▲ Figure 17.6

Note that, as shown above, a heterozygous individual will not be pure breeding.

A cross between a heterozygous tall pea plant and a dwarf pea plant

The heterozygous tall pea plant must be **Tt**. The dwarf pea plant must be **tt** (Figure 17.7).

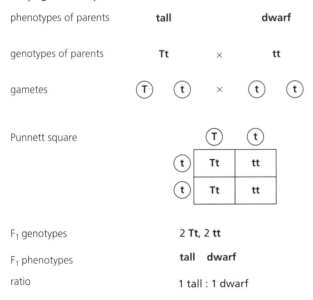

phenotypes of parents	**tall**		**dwarf**

genotypes of parents	**Tt**	×	**tt**

gametes (T) (t) × (t) (t)

Punnett square

	(T)	(t)
(t)	**Tt**	**tt**
(t)	**Tt**	**tt**

F_1 genotypes — 2 **Tt**, 2 **tt**

F_1 phenotypes — **tall   dwarf**

ratio — 1 tall : 1 dwarf

▲ Figure 17.7

In exam questions involving genetic crosses, you often need to predict the genotypes of the parents from descriptions of them. Remember that the dominant allele normally takes the capital letter of the characteristic it represents.

Pedigree diagrams and inheritance

The term pedigree often refers to the pure-breeding nature of animals, but is also used to describe human inheritance. Pedigree diagrams are similar to family trees and can be used to demonstrate how genetic disorders can be inherited. They can include symbols to indicate whether individuals are male or female, and what their genotype is for a particular genetic characteristic.

One genetic condition is called cystic fibrosis. Cystic fibrosis sufferers tend to have a much shorter lifespan than normal, and suffer from respiratory, digestive and reproductive problems.

● A person with cystic fibrosis has two recessive alleles (**ff**). A carrier of the condition has one normal allele and one recessive allele (**Ff**). A healthy person has two normal alleles (**FF**).

● A man who is not a carrier (**FF**) who has children with a woman who is not a carrier (**FF**) will produce 100% children who are not carriers (all **FF**).

● If one parent is a carrier for cystic fibrosis (**Ff**) and the other parent is not a carrier (**FF**), 50% of their children are likely to be carriers (**Ff**) and 50% will not be carriers (**FF**).

● However, if both parents are carriers, then the likely ratio of offspring of non-carriers/carriers/cystic fibrosis sufferers (**FF:Ff:ff**) is 1:2:1. So, there is a 1 in 4 chance of a child born to these parents having cystic fibrosis.

Figure 17.8 shows the inheritance of cystic fibrosis in a family.

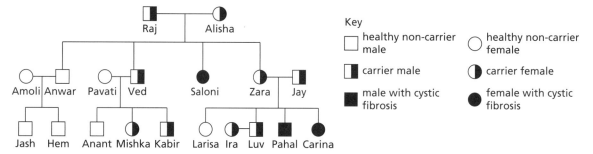

▲ **Figure 17.8 Pedigree diagram showing the inheritance of cystic fibrosis in a family**

Parents Raj and Alisha are married and both are cystic fibrosis carriers. However, because carriers have no symptoms of the disease, they may be unaware that they have defective alleles for cystic fibrosis. They go on to have four children. One child, Saloni, suffers from cystic fibrosis. The pedigree diagram shows that she does not get married and has no children. The three other children eventually get married and have children of their own.

The test-cross (back-cross)

A test-cross can be used to identify an unknown genotype. For example, a black mouse could have either the **BB** or the **Bb** genotype. One way to find out which it has is to cross the black mouse with a known homozygous recessive mouse (**bb**, having the phenotype of brown fur). The **bb** mouse will produce gametes with only the recessive **b** allele. A black homozygote (**BB**) will produce only **B** gametes.

● **BB × bb** will produce 100% black individuals (all **Bb**).

● **Bb × bb** will produce, on average, 50% black individuals (**Bb**) and 50% brown individuals (**bb**). This outcome identifies a parent that is not pure-breeding.

Codominance

This term is defined at the start of this chapter. **Codominance** describes a pair of alleles, neither of which is dominant over the other. This means that both can have an effect on the phenotype when they are present together in the genotype. The result is that there can be three different phenotypes. Therefore, codominance results in the appearance of a new characteristic that is intermediate to the parents' features. For example, if the parents are pure-breeding for long fur and short fur, the offspring will all have medium-length fur.

When writing the genotypes of codominant alleles, the common convention is to use a capital letter to represent the gene involved, and a small raised (superscript) letter for each phenotype.

Example

The alleles of the gene for flower colour in a plant are C^R (red) and C^W (white). The capital letter **C** has been chosen to represent 'colour'. Pure-breeding (homozygous) flowers may be red ($C^R C^R$) or white ($C^W C^W$). If these are cross-pollinated, all the first filial (F_1) generation will be heterozygous ($C^R C^W$), which are pink, because both alleles have an effect on the phenotype.

Self-pollinating the pink (F_1) plants results in a ratio in the next (F_2) generation of red:pink:white of 1:2:1.

Sample question   `REVISED`

The alleles of the gene for flower colour in a plant are C^R (red) and C^W (white). Pure-breeding (homozygous) flowers may be red ($C^R C^R$) or white ($C^W C^W$). If these are cross-pollinated, all the first filial (F_1) generation are pink because the alleles are codominant.

a State the genotype of the first filial (F_1) generation. [1]
b Write out a genetic cross as a Punnett square to show the results of self-pollinating the pink (F_1) plants. [3]
c Shade *all* the individuals associated with the Punnett square that are pink. [1]
d State the ratio of colours of flowers produced from the cross. [1]

Student's answer

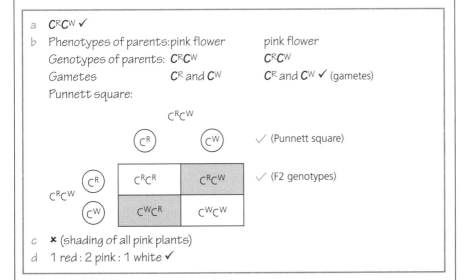

a $C^R C^W$ ✓
b Phenotypes of parents: pink flower   pink flower
    Genotypes of parents: $C^R C^W$ $C^R C^W$
    Gametes C^R and C^W C^R and C^W ✓ (gametes)
    Punnett square:

	C^R	C^W
C^R	$C^R C^R$	$C^R C^W$
C^W	$C^W C^R$	$C^W C^W$

✓ (Punnett square)
✓ (F2 genotypes)

c ✗ (shading of all pink plants)
d 1 red : 2 pink : 1 white ✓

Teacher's comments

The student has completed this question well, gaining 5 out of the 6 marks available. In part b there were marks for construction of the Punnett square, identification of the gametes and identification of the F_2 genotypes. The student had identified the gametes above the square correctly. The only mark lost was in part c, where the student overlooked the fact that parents involved in the cross are both pink, as well as two of the F_2 organisms.

Inheritance of the human ABO blood groups

These blood groups give an example of codominance. Instead of two alleles being present, in this case there are three: I^A, I^B and I^O. Combinations of these can result in four different phenotypes: A, B, AB and O. The alleles are responsible for producing antigens that respond to foreign antibodies (this can result in blood clotting in blood transfusions, and rejection of organs after transplant operations).

However, while I^A and I^B are codominant, I^O is dominated by both the other alleles. This means, for example, that a person with blood group A could have the genotype I^AI^A or I^AI^O. This has implications when having children because, if both parents carry the I^O allele, a child could be born with the genotype I^OI^O (blood group O), even though neither of the parents has this phenotype.

Example

Two parents have blood groups A and B. The father is I^AI^O and the mother is I^BI^O (Figure 17.9).

phenotypes of parents	**blood group A**		**blood group B**	
genotypes of parents	I^AI^O	×	I^BI^O	
gametes	I^A I^O	×	I^B I^O	

Punnett square

	I^A	I^O
I^B	I^AI^B	I^BI^O
I^O	I^AI^O	I^OI^O

F_1 genotypes	I^AI^O, I^BI^O, I^AI^B, I^OI^O
F_1 phenotypes	**A B AB O**
ratio	1 : 1 : 1 : 1

▲ **Figure 17.9**

Sex linkage

The definition of a **sex-linked characteristic** is given at the start of this chapter.

The sex chromosomes (X and Y) carry genes that control sexual development, but they also carry genes that control other characteristics These tend to be on the X chromosome, which has longer arms to the chromatids. Even if the allele is recessive, because there is no corresponding allele on the Y chromosome, it is bound to be expressed in a male (XY). There is less chance of a recessive allele being expressed in a female (XX) because the other X chromosome may carry the dominant form of the allele.

One example of this is red-green colour blindness (Figure 17.10). In the following case, the mother is a carrier of colour blindness ($X^R X^r$). This means that she shows no symptoms of colour blindness, but the recessive allele causing red-green colour blindness is present on one of her X chromosomes. The father has normal colour vision ($X^R Y$).

If the gene responsible for a particular condition is present on only the Y chromosome, only males can suffer from the condition because females do not possess the Y chromosome.

phenotypes of parents	mother: normal vision	father: normal vision

genotypes of parents $\quad X^R X^r \quad \times \quad X^R Y$

gametes $\quad (X^R) \ (X^r) \quad \times \quad (X^R) \ (Y)$

Punnett square

$X^R X^r$

	X^R	X^r
X^R	$X^R X^R$	$X^R X^r$
Y	$X^R Y$	$X^r Y$

$X^R Y$

F_1 genotypes $\quad X^R X^R \quad X^R X^r \quad X^R Y \quad X^r Y$

F_1 phenotypes $\quad$ 2 females with normal vision; 2 males, one with normal vision, one with red-green colour blindness

▲ **Figure 17.10**

Exam-style questions

1 a Distinguish between a gene and an allele. [2]
   b Draw a diagram to show that the inheritance of sex in humans produces 50% males and 50% females. [4]

2 Complete the following table to identify whether each of the following statements about DNA and protein formation is true or false. [8]

	Statement	True/false
a	The four bases present in DNA are A, C, G and P	
b	Each group of four bases codes for one amino acid	
c	A gene codes for one amino acid molecule	

	Statement	True/false
d	To make proteins, DNA must pass through the nuclear membrane	
e	mRNA instructs mitochondria to make a protein	
f	mRNA is a copy of a gene	
g	One of the genes on a DNA molecule will code for a fatty acid	
h	The gene for making a protein containing 56 amino acids will be made up of 168 bases	

3 a The nucleus of a human liver cell contains 46 chromosomes. Copy and complete the table below to show how many chromosomes would be present in the cells listed. [3]

Type of cell	Number of chromosomes
Ciliated cell in windpipe	
Red blood cell	
Ovum	

b Describe two differences, other than the number of chromosomes, between nuclei produced by mitosis and those produced by meiosis. [2]

4 Distinguish between:
a genotype and phenotype [2]
b homozygous and heterozygous [2]
c dominant and recessive [2]

5 Peas can be round or wrinkled, with round being dominant to wrinkled. State the genotype of each of the pea seeds described below.
a a heterozygous round pea [1]
b a wrinkled pea [1]
c a pure-breeding round pea [1]

6 Copy and complete the passage by writing the most appropriate word from the list in each space.

chromosome   diploid   gene   heterozygous   meiosis
mutation   phenotype   recessive   dominant

Petal colour in pea plants is controlled by a single _____ that

has two forms, red and white. The pollen grains are produced by

_____. After pollination, fertilisation occurs and the gametes

join to form a _____ zygote.

When two red-flowered pea plants were crossed with each other,

some of the offspring had white flowers. The _____ of the rest

of the offspring was red flowers. The white-flowered form is

_____ to the red-flowered form and each of the parent plants

was therefore _____. [6]

7 Explain why a male cannot be a carrier of red-green colour blindness. [2]

18 Variation and selection

Key objectives

The objectives for this chapter are to revise:
- definitions of the key terms
- continuous and discontinuous variation
- investigations into continuous and discontinuous variation
- that mutation is the way in which new alleles are formed
- factors that can increase the rate of mutation
- adaptive features and how to interpret them from images
- natural selection and selective breeding

- how selective breeding by artificial selection is used to improve crop plants and domesticated animals

- the sources of genetic variation in populations
- how to explain the adaptive features of hydrophytes and xerophytes
- the development of strains of antibiotic-resistant bacteria
- the difference between natural and artificial selection

Key terms

REVISED

Term	Definition
Adaptive feature	An inherited feature that helps an organism to survive and reproduce in its environment
Mutation	Genetic change
Variation	The differences between individuals of the same species
Adaptation	The process, resulting from natural selection, by which populations become more suited to their environment over many generations
Gene mutation	A random change in the base sequence of DNA

Variation

REVISED

The term **variation** is defined above. Those variations that can be inherited are determined by genes. They are **genetic variations**. There are two main types of variation: continuous and discontinuous.

Continuous variation

Continuous variation shows a complete range of phenotypes within a population, between two extremes. It is caused both by genes (often a number of different genes) and by the environment. Environmental influences for plants may be the availability of, or competition for, nutrients, light and water, and exposure to disease. For animals, environmental influences can include the availability of food or a balanced diet, and exposure to disease (or the availability of health services for humans).

Examples of continuous variation include height or body length, body mass and intelligence. When the frequency is plotted on a graph, as in Figure 18.1, a smooth curve is produced, with the majority of the population sample grouped together and only small numbers at the extremes of the graph.

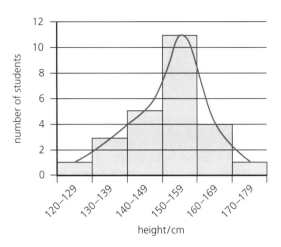

▲ Figure 18.1 Continuous variation in height

When plotting a graph of continuous variation, the bars should not have a gap between them. This is because it shows one characteristic with a range of numerical values. This type of graph is called a histogram.

Discontinuous variation

Discontinuous variation is seen where there is a limited number of obvious phenotypes for a feature, giving distinct categories. There are no intermediates between categories, and the feature cannot usually change during life. It is caused by a single gene or a small number of genes, with no environmental influence (as in Figure 18.2).

Examples include blood group, ability to tongue-roll and earlobe shape. When the frequencies are plotted on a graph, bars are produced that cannot be linked with a smooth curve.

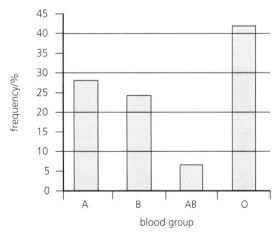

▲ Figure 18.2 Discontinuous variation in blood groups

When plotting a graph of discontinuous variation, the bars should have a gap between them. This is because the bars represent distinct categories. This is called a bar graph.

Sample question

REVISED

Seventy seeds were collected from a cross between two plants of the same species. The seeds were sown at the same time and, after 3 weeks, the heights of the plants that grew were measured and found to fall into two groups, A and B, as shown in Figure 18.3.

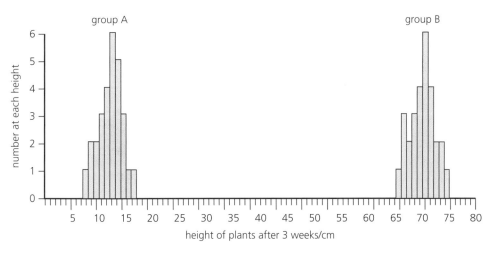

▲ **Figure 18.3**

a Calculate the percentage of seeds that germinated. Show your working. [2]
b i Name the type of variation shown within each group. [1]
ii State three factors that might have caused this variation. [3]

Student's answer

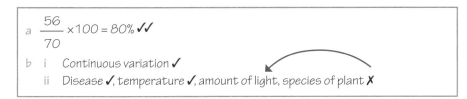

a $\dfrac{56}{70} \times 100 = 80\%$ ✓✓

b i Continuous variation ✓
ii Disease ✓, temperature ✓, amount of light, species of plant ✗

Teacher's comments

The answers for parts a and b(i) were good. However, there were 3 marks for b(ii) and the student gave four answers. As the last answer was incorrect, the third mark was not awarded.

Mutations

Mutations are a source of variation caused by an unpredictable change in the genes or chromosome numbers. As a result, new alleles are formed.

Mutations are normally very rare. However, exposure to ionising radiation and some chemicals, such as tar in tobacco smoke, increases the rate of mutation. Exposure can cause uncontrolled cell division, leading to the formation of tumours (cancer). The exposure of gonads (testes and ovaries) to radiation can lead to sterility or damage to genes in sex cells, which can be passed on to children.

Gene mutations

A **gene mutation** (see definition at the start of this chapter) can result in a genetic change. The sequence of bases in DNA becomes altered, resulting in a change in coding for one or more amino acids (see Chapter 4). A section of DNA may now start making a different protein that could affect the organism.

Sources of genetic variation in populations include mutation, meiosis (see Chapter 17), random mating, and random fertilisation.

Adaptive features

REVISED

The term **adaptive feature** is defined at the start of this chapter. You need to be able to interpret images and other information about a species to describe its adaptive features. Figure 18.4 gives an example.

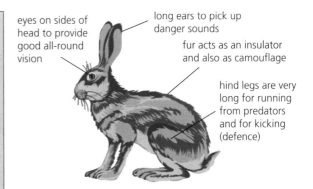

▲ **Figure 18.4 Adaptive features of a hare**

eyes on sides of head to provide good all-round vision

long ears to pick up danger sounds

fur acts as an insulator and also as camouflage

hind legs are very long for running from predators and for kicking (defence)

Skills

Identifying adaptations

● If you are given an image of an organism and asked to identify adaptations in it, try to compare it with a typical organism of the same group.

● Look at its key features – for example, are any features bigger or a different shape or missing?

● To practise this skill, print or copy a photograph or diagram of an organism (animal or plant) and annotate it to identify and describe the organism's adaptive features. An example (a hare) is shown in Figure 18.4.

● In your description, always state the adaptation and how it helps the organism to survive.

● Only describe adaptations you can actually see in the image.

Adaptations to arid conditions: xerophytes

Where possible, you should be able to describe these features based on plants you are familiar with and that grow in your local area. Some plants are adapted to cope with a lack of water (e.g. in very dry or 'arid' environments). These are called **xerophytes**.

▼ **Table 18.1 Adaptations of xerophytes**

Plant	Modifications
Pinus (pine tree)	Leaves are needle-shaped to reduce surface area for transpiration and to resist wind damage
	Sunken stomata to create high humidity and reduce transpiration
	Thick waxy cuticle on the epidermis to prevent evaporation from leaf surface

Adaptations to living in water: hydrophytes

Some plants are adapted to cope with living in water (e.g. pond plants and seaweeds). These are called **hydrophytes**.

▼ **Table 18.2 Adaptations of hydrophytes**

Plant	Modifications
Nymphaea (water lily)	Roots are poorly developed and contain air spaces (because of poor oxygen levels in the mud they grow in)
	Leaves contain large air spaces to make them buoyant, so they float on or near the surface (to gain light for photosynthesis)
	Lower epidermis of the leaves lack stomata to avoid waterlogging, while the upper epidermis has stomata for gas exchange
	Stems lack much support as the water surrounding them provides buoyancy for the plant

Selection

REVISED

Natural selection

Variation describes differences in a population. Some variation is genetic, which is passed on from parents. Animals and plants produced by sexual reproduction will show variation from their parents, for example in the size of the muscles in the legs of lions.

When organisms reproduce, many offspring are often produced. However, not all of them are likely to survive because of competition for resources such as food, water and shelter. The same is true for plants (they compete for resources such as nutrients, light, water and space). There is a struggle for survival.

The individuals with the most favourable characteristics are most likely to survive because they have an advantage over others in the population. For example, a lion cub with bigger muscles in its legs would be able to run more quickly and get food more successfully than its siblings.

In an environment where there is a food shortage, the individual with the best adaptations to the environment is most likely to survive to adulthood. The weaker individuals die before having the chance to breed, but the surviving adults breed and pass on the advantageous alleles to their offspring. More of the next generation carry the advantageous genes, resulting in a stronger population, better adapted to their environment.

Revision activity

Make a flowchart to identify the stages involved in the process of natural selection.

Adaptation

The definition of the process of **adaptation** is given at the start of this chapter. Slow changes in the environment result in adaptations in a population to cope with the change. Failure to adapt could result in the species becoming extinct. Over many generations, this gradual process results in populations becoming more suited to their environment.

Antibiotic-resistant bacteria

This is an example of natural selection. Bacteria reproduce rapidly – a new generation can be produced every 20 minutes by binary fission (see Chapter 16). Antibiotics are used to treat bacterial infections. An antibiotic is a chemical that kills bacteria by preventing bacterial cell wall formation.

Mutations occur during reproduction, which produce some variation in the population of bacteria. Individual bacteria with the most favourable features are most likely to survive and reproduce. A mutation may occur that enables a bacterium to resist being killed by an antibiotic treatment, while the rest of the population is killed when treated. This bacterium would survive the treatment and reproduce, passing on the antibiotic resistance gene to its offspring. Future treatment of this population of bacteria using the antibiotic would be ineffective.

Selective breeding (artificial selection)

Selective breeding is used by humans to produce varieties of animals and plants that have increased economic importance:

- Humans first select individuals with desirable features.
- These individuals are cross-bred to produce the next generation.
- From that generation, the offspring with desirable features are selected for further breeding.

Examples of improving crop plants and domesticated animals by selective breeding

Wild varieties of plants sometimes have increased resistance to fungal diseases, but have poor fruit yield. Cross-breeding wheat plants over a number of generations and selecting the organisms with the best features at each stage can result in the formation of varieties that have both high resistance to disease and high seed yield.

A variety of cattle may have a higher-than-average milk yield. Another variety may have a very high meat yield. If the two varieties are cross-bred, the individuals in the next generation with the best features can be selected to continue breeding until a new breed has been artificially produced with the benefits of both parental varieties (high milk production in females; high meat yield in males).

Comparing natural and artificial selection

Natural selection occurs in groups of living organisms through the passing on of genes to the next generation by the best-adapted organisms, without human interference. Those with genes that provide an advantage to cope with changes in environmental conditions are more likely to survive, while others die before they can breed and pass on their genes. However, variation within the population remains.

Artificial selection is used by humans to produce varieties of animals and plants that have an increased economic importance. It is considered a safe way of developing new strains of organisms and is a much faster process than natural selection. However, selective breeding removes variation from a population, leaving it susceptible to disease and unable to cope with changes in environmental conditions. So, potentially, artificial selection puts a species at risk of extinction.

Exam-style questions

1 a Distinguish between discontinuous and continuous variation. [2]
   b Describe the shape of a graph for an example of continuous variation. [2]

2 a Define the term *gene mutation*. [2]
   b State three sources of genetic variation in a population. [3]

3 Figure 18.5 shows a polar bear and a sun bear. The polar bear lives in a very cold climate, often moving on snow and ice. The sun bear lives in a more favourable climate.

▲ **Figure 18.5**

   a Define the term *adaptive feature*. [2]
   b Complete the following table by:
      i   stating two ways, visible in Figure 18.5, in which the polar bear is adapted for life in a cold climate [2]
      ii giving one advantage of each of the adaptations you have identified [2]

	Adaptation	Advantage
1		
2		

4 a State the name given to a plant that is adapted for living in water. [1]
   b What are the advantages to a water plant of having:
      i   large air spaces in its leaves [2]
      ii   no stomata on the lower epidermis of the leaves [2]
      iii roots containing air spaces? [2]

5 Farmers have carried out selective breeding to improve the breeds of some animals. Some of the original breeds have become very rare and are in danger of becoming extinct.
   a Explain the meaning of selective breeding. [2]
   b Outline how selective breeding has been used to develop a named variety of animal or plant. State the characteristics of the new variety. [4]

Key objectives

The objectives for this chapter are to revise:
- definitions of the key terms
- the role of the Sun in biological systems
- the flow of energy between organisms and its transfer to the environment
- how to construct and interpret simple food chains
- how to use food chains and food webs to describe the impacts humans have
- how to draw, describe and interpret pyramids of numbers and biomass
- how to identify and name the trophic levels in food chains, food webs and ecological pyramids
- the carbon cycle
- the effects of burning fossil fuels and cutting down forests on the carbon cycle
- the factors affecting the rate of population growth for a population of an organism
- how to identify the phases of a sigmoid population growth curve
- how to interpret graphs and diagrams of population growth

- how to draw, describe and interpret pyramids of energy
- the advantages of using a pyramid of energy
- why the transfer of energy from one trophic level to another is often not efficient
- why food chains usually have fewer than five trophic levels
- why it is more energy efficient for humans to eat crop plants
- the nitrogen cycle and the roles of microorganisms in it
- explanations for the factors that lead to each phase of a sigmoid population growth curve, and the role of limiting factors

Key terms

Term	Definition
Carnivore	An animal that gets its energy by eating other animals
Community	All of the populations of different species in an ecosystem
Consumer	An organism that gets its energy by feeding on other organisms
Decomposer	An organism that gets its energy from dead or waste organic material
Ecosystem	A unit containing the community of organisms and their environment, interacting together
Food chain	A chart showing the transfer of energy from one organism to the next, beginning with a producer
Food web	A network of interconnected food chains
Herbivore	An animal that gets its energy by eating plants
Population	A group of organisms of one species, living in the same area, at the same time
Producer	An organism that makes its organic nutrients, usually using energy from sunlight, through photosynthesis
Trophic level	The position of an organism in a food chain, food web or ecological pyramid

Revision activity

This section of the syllabus has a large number of key terms and their definitions, which you need to learn. First, make a set of cards, with one card for each key term. Then make another set of cards, with one definition on each card. Mix up the cards. Then try to match each of the terms with its definition.

Energy flow

The Sun is the principal source of energy input to biological systems. The Earth receives two main types of energy from the Sun: light (solar) and heat. Photosynthetic plants and some bacteria can trap light energy and convert it into chemical energy.

Heterotrophic organisms obtain their energy by eating plants or animals that have eaten plants. So, all organisms, directly or indirectly, get their energy from the Sun. The chemical energy produced is passed from one organism to another in a food chain but, unlike water and elements such as carbon and nitrogen, energy does not return in a cycle. The energy given out by organisms is lost to the environment.

Food chains and food webs

Food chains

The definition of a **food chain** is given at the start of this chapter. Energy is transferred between organisms in a food chain by **ingestion**. Food chains are lists of organisms that show the feeding relationship between them, as in the example below:

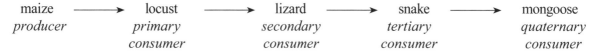

maize	locust	lizard	snake	mongoose
producer	*primary consumer*	*secondary consumer*	*tertiary consumer*	*quaternary consumer*

A food chain usually starts with a **producer** (a photosynthetic plant), which gains its energy from the Sun. The Sun itself should not be included, as it is not an organism. The arrows used to link each organism to the next represent the transfer of energy. They always point towards the 'eater' and away from the plant. The feeding level is known as the **trophic level**.

- Plants are producers (they make, or produce, food for other organisms).
- Animals that eat plants are primary **consumers** (a consumer is an 'eater'). They are also called **herbivores**.
- Animals that eat other animals are secondary, or possibly tertiary, consumers, depending on their position in the chain. They are also called **carnivores**.
- When an organism dies, it may be broken down by a **decomposer**. Decomposers are organisms that get their energy from breaking down dead or waste organic material. Conditions need to be suitable for this to happen – for example, temperature, availability of oxygen and water.

Skills

Constructing food chains
You need to be able to draw a food chain involving at least three consumers:
- Always start with the producer on the left of the diagram.
- Make sure that the arrows are pointing in the correct direction.
- Practise labelling each trophic level in your food chain under the organisms (producer, primary consumer etc.)
- Do not waste time drawing the plants and animals – this will not get you any extra marks.

Food webs

The definition of a **food web** is given at the start of this chapter. Food webs are a more accurate way of showing feeding relationships than food chains, because most animals have more than one food source.

For example, in the food web in Figure 19.1, the leopard feeds on both baboons and impala.

The producer is grass. Locusts and impala are both primary consumers (herbivores), the scorpion is a secondary consumer (carnivore) and the baboon is a tertiary consumer (carnivore). The leopard is acting as both a secondary and a quaternary (fourth) consumer.

Food chains and webs are easily unbalanced, especially if one population of organisms in the web dies or disappears. Humans can be the cause, through over-harvesting (perhaps by over-predation or hunting) or through the introduction of foreign species to a habitat.

For example, in the food web in Figure 19.1, if all the baboons were killed by hunters, the leopard would have only impala to eat, and so the impala population would decrease. The scorpion population may increase because of less predation by baboons but, if there are more scorpions, they will eat more locusts, reducing the locust population, and so on.

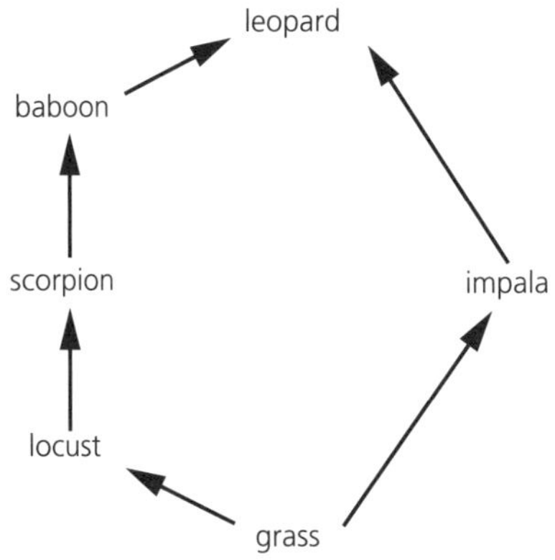

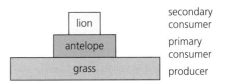

▲ **Figure 19.1 A food web**

Pyramids of numbers

In a pyramid of numbers, each trophic level in a food chain is represented by a horizontal bar, with the width of the bar representing the number of organisms at that level. The base of the pyramid represents the producer, the second level is the primary consumer, and so on. Figure 19.2 shows a typical pyramid of numbers.

Usually, the producers have the largest numbers, so they form the widest bar. There will be fewer primary consumers, and even fewer secondary consumers, so a pyramid shape is formed. However, this is not always true. Figure 19.3 shows a different pyramid of numbers.

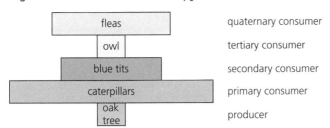

▲ **Figure 19.2 A typical pyramid of numbers**

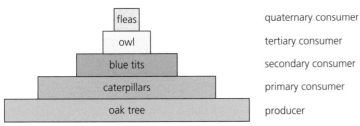

▲ **Figure 19.3 Pyramids of numbers do not always form a 'pyramid'**

The food chain for Figure 19.3 is supported by a single organism (a large oak tree). Many caterpillars feed on its leaves. Only a single owl is supported by the blue tits. However, the owl has many fleas, which feed on it by sucking its blood.

Pyramids of biomass

Figure 19.3 shows that pyramids of numbers are limited in what they show. It is more useful to measure the amount of living material (biomass) at each level over a fixed area of habitat. Once this is done, a normal-shaped pyramid is usually obtained, as shown in Figure 19.4.

fleas — quaternary consumer
owl — tertiary consumer
blue tits — secondary consumer
caterpillars — primary consumer
oak tree — producer

▲ **Figure 19.4 A pyramid of biomass**

You need to be able to identify the trophic levels in food chains, food webs, pyramids of numbers and pyramids of biomass. Sometimes (as in Figure 19.1) there is a fourth level of consumer. This is called a **quaternary consumer**.

Pyramids of energy

A pyramid of energy is a graphic representation of the flow of energy through each trophic level in a food chain over a fixed period of time. Data for this sort of display are produced by calculating the energy available over a fixed period of time, for example a year, in each trophic level. They are shown as bars in a pyramid.

For example, the energy available in a year's supply of leaves is compared with the energy needed to maintain a population of insects that feed on the leaves. The producers at the base of the pyramid will have the greatest amount of energy and each successive trophic level would show a reduced amount of energy.

Advantages of pyramids of energy over other food pyramids

- They are never inverted.
- They take account of the rate of production over a period of time, so they can be compared with other energy pyramids.
- The amount of solar energy entering at the producer level can be taken into account.
- They can also be constructed using data for all the organisms involved in food webs, or for whole communities in an ecosystem.
- They show the efficiency of energy transfer.

Energy transfer

Energy is lost at each level in the food chain, as it is transferred between trophic levels. The following examples show how the energy is lost:

- Energy lost through the process of respiration (as heat).
- Energy used up for movement (to search for food, find a mate, escape from predators etc.).
- Warm-blooded animals (birds and mammals) maintain a constant body temperature – they lose heat to the environment.
- Warm-blooded animals lose heat energy in faeces and urine.
- Some of the material in the organism being eaten is not used by the consumer; for example, a locust does not eat the roots of maize, and some of the parts eaten are not digestible.

Even plants do not make use of all the light energy available to them. This is because some light:

- is reflected off shiny leaves
- is the wrong wavelength for chlorophyll to trap
- passes through the leaves without meeting any chloroplasts
- does not fall on the leaves

This means that the transfer of energy between trophic levels is inefficient – a lot is lost. On average, about 90% of the energy is lost at each level in a food chain. This means that, in long food chains, very little of the energy entering the chain through the producer is available to the top carnivore. Thus, there tend to be small numbers of top carnivores, and food chains usually have fewer than five trophic levels.

The food chain below shows how energy reduces through the chain. It is based on maize obtaining 100 units of energy:

maize	→	locust	→	lizard	→	snake
100 units		10 units		1 unit		0.1 unit

In shorter food chains, less energy is lost. In terms of conservation of energy, short food chains are more efficient than long ones in providing energy to the top consumer. Below are two food chains and the energy values for each level in them. Both food chains have a human being as the top consumer:

maize	→	cow	→	human
100 units		10 units		1 unit

maize	→	human
100 units		10 units

Ten times more energy is available to the human in the second food chain than in the first. In the second food chain, the human is a herbivore (vegetarian).

Some farmers try to maximise meat production by reducing movement of their animals (keeping them in pens or cages with a food supply), and keeping them warm in winter. This means less stored energy is wasted by the animals.

Nutrient cycles

The carbon cycle

Figure 19.5 shows the main parts of the carbon cycle.

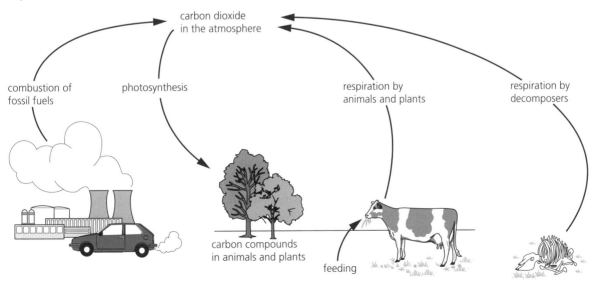

▲ **Figure 19.5 The carbon cycle**

Diagrams of the carbon cycle may look complicated, but only five main processes are involved: photosynthesis, respiration, decomposition, fossilisation and combustion:

- Carbon moves into and out of the atmosphere mainly in the form of carbon dioxide.
- Plants take carbon dioxide out of the air during photosynthesis.
- Plants convert carbon dioxide into organic materials (carbohydrates, fats and proteins).
- Herbivores obtain carbon compounds by feeding on plants. Carnivores gain carbon compounds by feeding on other animals.
- Animals and plants release carbon dioxide back into the air through respiration.
- When organisms die, they usually rot (the process of decomposition). Decomposers break down the organic molecules through the process of respiration to release energy. This also releases carbon dioxide into the air.
- If a dead organism does not decompose, the carbon compounds are trapped in its body. Over a long period, this can form fossil fuels, such as coal, oil or gas (fossilisation).
- Combustion of fossil fuels releases carbon dioxide back into the air.

If there is an increase in the combustion of fossil fuels, or if more trees are cut down and not replaced, carbon dioxide levels in the atmosphere will increase. This contributes to global warming. Carbon dioxide forms a layer in the atmosphere, which traps heat radiation from the Sun. This causes a gradual increase in the atmospheric temperature, which can:

- melt polar ice caps, causing flooding of low-lying land
- change weather conditions in some countries, increasing flooding or reducing rainfall and changing arable (farm) land to desert
- cause the extinction of some species that cannot survive at higher temperatures

Revision activity

Make flash cards for each of the processes involved in the carbon cycle.

For each process, describe what is happening to carbon.

The nitrogen cycle

Figure 19.6 shows the main parts of the nitrogen cycle. You do not need to know the names of individual bacteria, but you do need to know the roles of the three main types:

- Nitrogen-fixing bacteria – convert nitrogen gas into compounds of ammonia.
- Nitrifying bacteria – convert compounds of ammonia into nitrates.
- Denitrifying bacteria – break down nitrates into nitrogen gas.

The element nitrogen is a very unreactive gas. Plants are not able to change it into nitrogen compounds, but it is needed to form proteins. Nitrogen compounds become available for plants in the soil in a number of ways, including:

- nitrogen-fixing bacteria (some plants – legumes such as peas, beans and clover – have roots with nodules that contain these bacteria, so the plant receives a direct source of nitrates)
- breakdown of dead plants and animals by decomposers (bacteria, fungi and invertebrates)
- the addition of artificial fertilisers, compost (decaying plant material) and manure (decaying animal waste – urine and faeces)
- lightning – its energy causes nitrogen to react with oxygen

Plants absorb nitrates into their roots by active uptake (see Chapter 3). The nitrates are combined with glucose (from photosynthesis) to form amino acids and proteins. Proteins are passed through the food chain as animals eat the plants. When animals digest proteins, the amino acids released can be reorganised to form different proteins. Animals cannot store surplus amino acids. They are broken down in the liver in a process called deamination (see Chapter 13). This involves

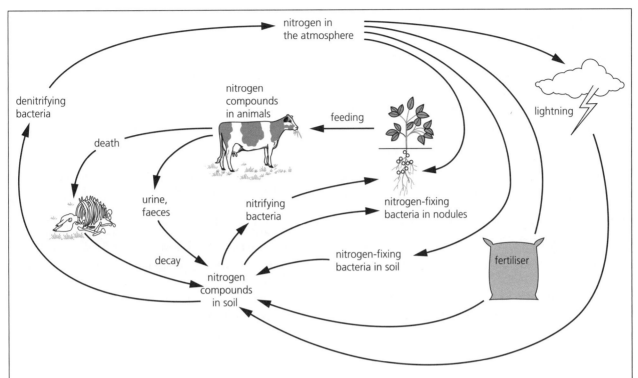

▲ **Figure 19.6 The nitrogen cycle**

removing the nitrogen-containing part of the amino acids to form urea.

Some soil bacteria – denitrifying bacteria – break down nitrogen compounds and release nitrogen back into the atmosphere. This is a destructive process, commonly occurring in waterlogged soil.

Farmers try to keep soil well drained to prevent this happening – a shortage of nitrates in the soil stunts the growth of crop plants.

Nitrates and other ammonium compounds are very soluble, so they are easily leached out of the soil and can cause pollution (see Chapter 20).

Sample question

Figure 19.7 shows the nitrogen cycle.

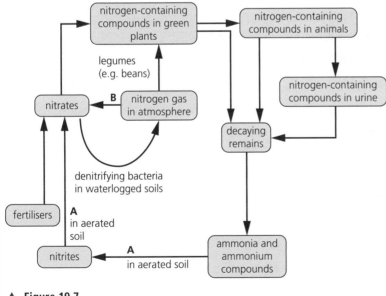

▲ **Figure 19.7**

a i Name the main nitrogen-containing compound found both in plants and in animals. [1]
  ii Name one nitrogen-containing compound that is present in urine. [1]
  iii Name the type of organism that causes the changes at A. [1]
  iv What atmospheric conditions bring about the change at B? [1]
b Using the figure, explain why it is an advantage to have good drainage in most agricultural land. [4]

Student's answer

a i Protein ✓
  ii Urea ✓
  iii Microbes ✗
  iv Lightning ✓
b Waterlogged soils contain denitrifying bacteria ✓, which change nitrates to nitrogen gas ✓. Most plants cannot use nitrogen and they would be short of nitrates for them to absorb. ✓

Teacher's comments

The answers to part a have gained 3 out of the 4 marks available. 'Microbes' is too general for part a(iii) – *bacteria* was the response needed. The response to part b is good, but a fourth mark was available for stating what plants use the nitrate for (formation of proteins) or the effect of a shortage of nitrates on the plant (poor growth).

Populations

REVISED

You need to be able to define the terms **community** and **ecosystem**. The terms **population**, community and ecosystem are associated, as shown in Figure 19.8.

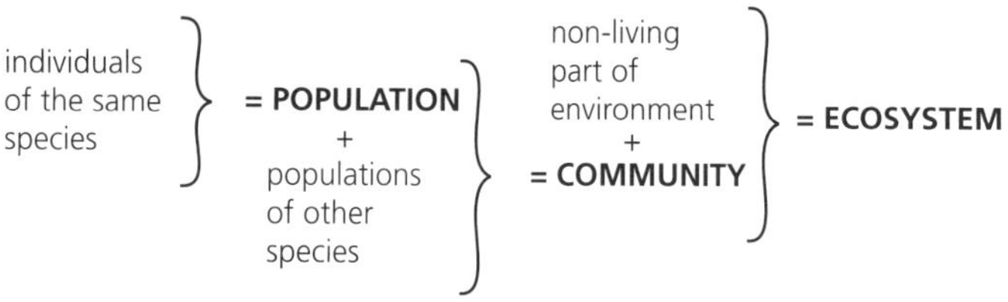

individuals of the same species } = **POPULATION** + populations of other species } non-living part of environment + = **COMMUNITY** } = **ECOSYSTEM**

▲ Figure 19.8 The association between population, community and ecosystem

In a lake, for example, the animal **community** will include **populations** of fish, insects, crustaceans, molluscs and protoctists. The plant community will consist of rooted plants with submerged leaves, rooted plants with floating leaves, reed-like plants growing at the lake margin, plants floating freely on the surface and algae in the surface waters.

A lake is an **ecosystem**, which consists of the plant and animal communities mentioned above, and the non-living part of the environment (mud, water, minerals, dissolved oxygen, soil and sunlight) on which they depend.

Factors affecting population growth

The rate of growth of a population depends on the following:

- **Food supply** – ample food will enable organisms to breed more successfully to produce more offspring; a shortage of food can result in death or can force emigration, reducing the population.
- **Competition** – within a habitat there will be competition for factors such as food and shelter. This competition may be between individuals of the same species, or between individuals of different species if they

eat the same food. There will also be competition within a species for mates.

- **Predation** – if there is heavy predation of a population, the breeding rate may not be sufficient to produce enough organisms to replace those eaten, so the population will drop in numbers. There tends to be a time lag in population size change for predators and their prey. As predator numbers increase, prey numbers drop, and as predator numbers drop, prey numbers rise again (unless there are other limiting factors).
- **Disease** – this is a particular problem in large populations, because disease can spread easily from one individual to another. Epidemics can reduce population sizes very rapidly.

Sigmoid population growth curves

A population growing in an environment with limited resources does not produce a straight line when plotted on a graph. Instead, a sigmoid (S-shaped) curve is formed, as shown in Figure 19.9 for a colony of yeast.

You need to be able to identify the **lag**, exponential (**log**), **stationary** and **death** phases on a graph of population growth.

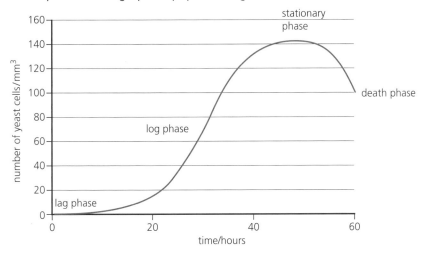

▲ **Figure 19.9 Graph of population growth**

If you need write these terms in the exam, make sure that your letters – 'o' in 'log' and 'a' in 'lag' – are clear and recognisable.

You need to be able to explain the factors that lead to the different phases shown in Figure 19.9:

- **Lag phase** – the new population takes time to settle and mature before breeding begins. When this happens, a doubling of small numbers does not have a big impact on the total population size, so the line of the graph rises only slowly with time.
- **Log (exponential) phase** – there are no limiting factors. Rapid breeding in an increasing population causes a significant increase in numbers. A steady doubling in numbers per unit of time produces a straight line.
- **Stationary phase** – limiting factors, such as shortages of food, cause the rate of reproduction to slow down and there are more deaths in the population. When the birth rate and death rate are equal, the line of the graph becomes horizontal.

● **Death phase** – the mortality rate (death rate) is now greater than the reproduction rate, so the population numbers begin to drop. Fewer offspring will live long enough to reproduce. The decline in population numbers can happen because the food supply is insufficient, waste products contaminate the habitat or disease spreads through the population.

Limits to population growth

A limiting factor, such as food, takes effect as the population becomes too large for supplies to be sufficient. The population growth rate reduces until births and deaths are equal. At this point, there is no increase in numbers – the graph forms a plateau. As food runs out, more organisms die than are born, so the number in the population drops. This is the death phase.

Sample question

REVISED

Figure 19.10 shows changes in a population of small carnivores in a new habitat.

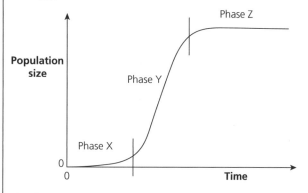

▲ **Figure 19.10**

a State the name given to this type of curve. [1]
b Identify phases X, Y and Z shown on the graph. [3]
c Suggest two reasons why phase Z becomes horizontal for these carnivores. [2]

Student's answer

a Sigmoid curve ✔
b X = lag phase ✔, Y = log phase ✔, Z = stationary phase ✔
c There are limiting factors. ✗ More deaths than births. ✗

Correct answer

a Sigmoid curve
b X = lag phase, Y = log phase, Z = stationary phase
c A horizontal line means that there are the same numbers of births and deaths. This may due to limiting factors, such as a lack of food.

Teacher's comments

The student started well, giving correct answers to parts a and b. However, neither of the suggestions given to explain why the stationary phase becomes horizontal were correct. The first (limiting factors) was too vague. Naming a limiting factor, such as lack of food, would have been acceptable. The second suggestion was incorrect: if there had been more deaths than births, the population numbers would start to drop rather than staying horizontal. A horizontal line means that the numbers of births and deaths are the same.

Exam-style questions

1 State the difference between the terms in each of the following pairs:
   a producer and consumer [4]
   b carnivore and herbivore [2]
2 The following diagram shows a food chain:

   pawpaw tree → flying ants → spiders → tree shrew → marbled cat
   a State the source of energy for the pawpaw tree. [1]

   b State two different biological terms that could be used to describe the position of the tree shrew in the food chain. [2]
   c Describe what the arrows represent in the food chain. [2]
   d State the term that describes the position of an organism in a food chain. [1]
   e Suggest, with reasons, the effect of a reduction in marbled cats on:
   i tree shrews [2]
   ii pawpaw trees [2]

3 Figure 19.11 shows a food web.

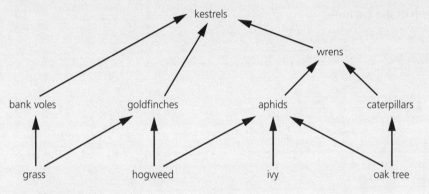

▲ **Figure 19.11**

   a Select appropriate organisms from the food web to complete each column in the table below. [4]

	Consumer	Producer	Carnivore	Herbivore
Organism 1				
Organism 2				

   b Ladybirds eat aphids. A very large number of ladybirds arrive in the habitat where these organisms live. Predict some of the possible effects this could have on the organisms in the above food web. [6]

4 Figure 19.12 shows the flow of energy through a complete food chain.

decomposers

34 000   3000   360   Y

Sun →
producer
50 000
first trophic level
→
first consumer
6000
second trophic level
→
second consumer
X
third trophic level
→
third consumer
240
fourth trophic level

10 000   1800   600   192

lost from food chain – mainly through respiration

numbers are energy values in arbitrary units

▲ **Figure 19.12**

a i Which form of the Sun's energy is trapped by the producer? [1]
  ii Into which energy form is the Sun's energy converted when it is trapped by the
     producer? [1]
b i The first consumer has received 6000 units of energy. How many units of energy
     (X on the figure) have been passed to the second consumer? [1]
  ii How many units of energy (Y on the figure) are lost from the third consumer to the
     decomposers? [1]
c i Suggest why the proportion of the energy intake that a producer loses to the
     environment (20%) is smaller than that lost to the environment by a first consumer (30%). [2]
  ii Many countries have difficulty in producing enough food for their population. If humans
     were always fed as first consumers, rather than as second or third consumers,
     how might this help to overcome this problem? [3]

5 Figure 19.13 shows a diagram of the carbon cycle.

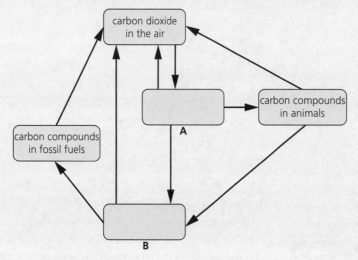

▲ Figure 19.13

a Copy and complete the cycle by filling in   i   combustion – C   [1]
  boxes A and B.   [2]   ii decomposition – D   [1]
b On your diagram, label with the letter   iii photosynthesis – P   [1]
  indicated an arrow that represents the   iv respiration – R   [1]
  process of:

6 Outline the role of bacteria in:
  a the carbon cycle [2]
  b the nitrogen cycle [6]
7 Figure 19.14 shows a population curve for a species of animal colonising a new habitat.

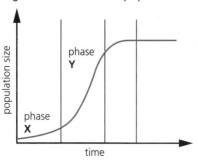

▲ Figure 19.14
a i Identify phase Y of the curve. [1]
  ii Suggest why the population increase in phase X is slow. [2]
b Identify three factors that limit the size of such a population, but do not appear
  to limit the total human population. [3]

Key objectives

The objectives for this chapter are to revise:
- definitions of key terms
- how modern technology has resulted in increased food production
- the positive and negative impacts on an ecosystem of large-scale monocultures of crop plants and of intensive livestock production
- the reasons for habitat destruction and the undesirable effects of deforestation on the environment
- the sources and effects of pollution of land, water and air
- the effects of increases in carbon dioxide and methane concentrations in the atmosphere

- that some resources can be conserved and maintained sustainably
- why organisms become endangered or extinct, and how endangered species can be conserved

- the process of eutrophication of water
- how forests and fish stocks can be conserved
- the reasons for conservation programmes
- the use of artificial insemination and *in vitro* fertilisation in captive breeding programmes
- the risks to a species if the population size drops, reducing genetic variation

Key terms

REVISED ☐

Term	Definition
Biodiversity	The number of different species that live in an area
Sustainable resource	A resource that is produced as rapidly as it is removed from the environment, so that it does not run out

Food supply

REVISED ☐

Larger populations require more food, provided by improving methods of agriculture. Modern technology has resulted in increased food production in a number of ways:

- **Agricultural machinery** enables much larger areas of land to be cleared, and makes preparing soil, and planting, maintaining and harvesting crops significantly more efficient. The process of farming in general has become more efficient.

- The use of **chemical fertilisers** improves yield. These are mineral ions made on an industrial scale. Examples are ammonium sulfate (for nitrogen and sulfur), ammonium nitrate (for nitrogen) and compound NPK fertiliser for nitrogen, phosphorus and potassium. These are spread on the soil in carefully calculated amounts to provide the minerals that the plants need.

- The use of **insecticides** improves quality and yield. Crops are very susceptible to attack by insect pests. Insecticides combat these attacks, so the crops grow more successfully and show less damage.

- The use of **herbicides** reduces competition with weeds. Weeds are plants that compete with the crop plant for root space, soil minerals and sunlight. Herbicides are chemicals that kill the weeds growing among the crop plants.

- **Selective breeding** can be used to improve production by crop plants and livestock. An important part of any breeding programme is the selection of the desired varieties that have particular qualities, such as flavour and disease resistance in plants, and high milk or meat yield or resistance to disease in animals such as cattle, fish and poultry.

Monoculture

A monoculture (Table 20.1) is a crop grown on the same land, year after year. The land is maintained so that all the organisms that feed on, compete with, or infect the crop plant are eradicated.

▼ Table 20.1 Advantages and disadvantages of large-scale monocultures of crop plants

Advantages	Disadvantages
Increases crop yield	The number of species in an area are reduced (reduced biodiversity)
Can be managed more efficiently with agricultural machinery	There is a negative impact on food chains
More crops can be grown on less land	The removal of hedges reduces nesting sites for birds and habitats for other organisms
Greater profits for the farmer	The use of pesticides on monocultures can reduce the number of important insect pollinators, which are required by wild flowers

Intensive livestock production

Intensive livestock production (Table 20.2) is also known as 'factory farming'. Chickens and calves are often reared in large sheds instead of in open fields.

▼ Table 20.2 Advantages and disadvantages of intensive livestock production

Advantages	Disadvantages
The yield is very high Many of the systems required to rear the animals can be automated, which is cheaper and less labour intensive Greater profits for the farmer	The animals' urine and faeces are washed out of the sheds with water, forming 'slurry'; if this gets into streams and rivers, it supplies an excess of nitrates and phosphates, which can lead to water pollution (eutrophication) Overgrazing can result from too many animals being kept on a pasture – they eat the grass down almost to the roots, and their hooves trample the surface soil into a hard layer; as a result, the rainwater will not penetrate the soil, so it runs off the surface, carrying the soil with it, and the soil becomes eroded. Being in confinement in large numbers can make animals more vulnerable to disease and to have less disease resistance

Habitat destruction

REVISED

Habitat destruction leads to a loss of **biodiversity**, which is defined at the beginning of the chapter. There are three key reasons for habitat destruction:

- An increased area of land is needed for food crop growth, livestock production and housing as the human population increases.
- As we need more raw materials for the manufacturing industry and increased energy supplies, there is more extraction of natural resources.
- Aquatic habitats are becoming contaminated with human debris, including untreated sewage, agricultural fertilisers, pesticides, non-biodegradable plastics and waste oil, leading to marine and freshwater pollution.

Food chains and food webs

If human activity causes one population of organisms to die or disappear, a food chain or food web becomes unbalanced. For example, an increase in herbivores due to the over-hunting of a carnivore may result in the overgrazing of land. Once the plants have been removed, the soil is vulnerable to erosion because there are no roots to absorb water or to hold the soil together. The habitat would then be destroyed.

Sample question

Figure 20.1 shows a zebra feeding in its natural habitat.

Zebra eats grass

▲ **Figure 20.1**

As part of a farming programme, a large herd of herbivore X was introduced to the habitat. Herbivore X breeds rapidly and feeds on roots by digging the plants up.

Describe and explain how the introduction of herbivore X might affect the ecosystem. [5]

Student's answer

> The population of zebra will get smaller ✔ and there will be fewer plants ✔.

Teacher's comments

The question contains two command words – 'describe' and 'explain'. This student has answered the question by describing the effect of introducing herbivore X but has not included an explanation. If the answer had been reworded to say 'There would be fewer plants because herbivore X has eaten the plant roots', this would be both a description and an explanation. An answer stating that 'As there are fewer plants, the zebra population would reduce because they feed on those plants' would also contain both a description and an explanation. The student could also have suggested how the habitat might have been affected, as this is part of an ecosystem – for example, 'Because there are fewer plant roots in the soil, soil erosion could occur'.

Correct answer

There will be fewer plants because herbivore X has eaten them and their roots. This will cause the population of zebra to shrink, as there will be less for them to feed on. The reduction in the number of plants might also lead to soil erosion because there are no roots to absorb water or to hold the soil together. Populations of carnivores may decrease if they feed on zebras but not herbivore X.

The undesirable effects of deforestation

Deforestation is the removal of large areas of forest to provide land for farming and roads, and to provide timber for building, furniture and fuel. The removal of large numbers of trees results in habitat destruction on a massive scale, which can have the follows results:

- Animals living in the forest lose their habitats and sources of food. Species of plant become extinct as the land is used for other purposes such as agriculture, mining, housing and roads, so biodiversity is reduced.
- Soil erosion is more likely to happen, because there are no roots to hold the soil in place. The soil can end up in rivers and lakes, destroying habitats there.
- Flooding becomes more frequent, as there is no soil to absorb and hold rainwater. Plant roots rot and animals drown, destroying food chains and webs.
- Carbon dioxide builds up in the atmosphere, because there are fewer trees to photosynthesise, increasing global warming. Climate change affects habitats.

Pollution

REVISED

Sewage

Untreated sewage contains disease organisms, which may get into drinking water and spread diseases such as typhoid and cholera. It also attracts vermin, which are vectors of disease.

Fertilisers

It is very tempting for farmers to increase the amount of fertilisers applied to crops to try to increase crop yields. However, this can lead to the eutrophication of rivers and lakes. Overuse of fertilisers can also lead to the death of the plants. High concentrations of fertiliser around plant roots can cause the roots to lose water by osmosis (see chapter 3). The plant then wilts and dies.

Eutrophication

Sewage and fertilisers both contain high levels of nutrients, such as nitrates and other ions. The nitrates act as fertilisers for producers, such as algae, which grow and die more rapidly. Decomposers such as bacteria feed on the dead organic matter and reproduce rapidly, using up dissolved oxygen as they respire aerobically. Animals in the water system die because of a lack of dissolved oxygen for aerobic respiration. Figure 20.2 shows a flowchart of eutrophication.

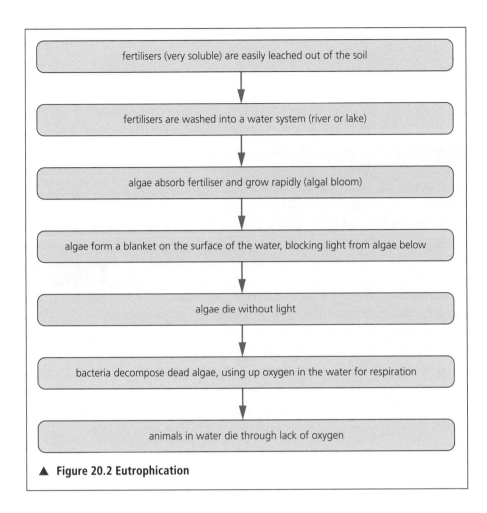

▲ **Figure 20.2 Eutrophication**

Discarded plastics

Non-biodegradable plastics are not broken down by decomposers when dumped in landfill sites or left as litter. This means that they remain in the environment, taking up valuable space or causing visual pollution. Discarded plastic bottles can trap small animals, and nylon fishing lines and nets can trap birds and mammals, such as seals and dolphins. As the plastic gradually breaks up into smaller fragments in the sea, it can clog up the gills of fish or get trapped in their stomachs, making them ill.

The greenhouse effect and climate change

Levels of **carbon dioxide** in the atmosphere are influenced by natural processes and by human activities. The main source of pollution that changes the equilibrium (balance) is the combustion of fossil fuels (coal, oil and gas). An increase in the levels of carbon dioxide in the atmosphere is thought to contribute to global warming. Carbon dioxide forms a layer in the atmosphere, which traps heat radiation from the Sun.

Methane also acts as a greenhouse gas. It is produced by the decay of organic matter in anaerobic conditions, such as in wet rice fields and in the stomachs of animals, such as cattle and termites. It is also released from the ground during the extraction of oil and coal.

The build-up of greenhouse gases causes a gradual increase in the atmospheric temperature, known as the **enhanced greenhouse effect**. This can:

- melt polar ice caps, causing flooding of low-lying land
- change weather conditions in some countries, by increasing flooding or reducing rainfall and thus changing arable (farm) land to desert; extreme weather conditions become more common
- cause the extinction of some species that cannot survive in raised temperatures

Revision activity

Make a spider diagram, with 'climate change' in the centre. On your diagram, include the causes of climate change and their effects on habitats and the species in them.

Conservation

REVISED

The definition of a **sustainable resource** is given at the start of this chapter. Some natural resources (the materials we take from the Earth) are non-renewable. For example, fossil fuels such as coal take millions of years to form. Increasing demands for energy are depleting these resources.

However, some resources, such as forests and fish stocks, can be conserved and maintained with careful management. This may involve replanting land with new seedlings as mature trees are felled, and controlling fishing activities where fish stocks are being depleted.

Conservation of forests

- **Education** – This usually involves sharing information with local communities about the need for conservation. Once they understand its importance, the environment they live in is more likely to be cared for and the species in it protected. In tropical rainforests, it has been found that the process of cutting down the trees damages twice as many trees next to them, and dragging the trees out of the forest creates more damage. Educating people about alternative ways of tree felling, reduction of wastage and the selection of species of trees to be felled makes the process more sustainable and helps to conserve rarer species.

- **Protected areas** – Areas where rare species of trees grow are protected to prevent them being felled for timber or to clear the land for other purposes. In this way the forests are being conserved.

- **Legal quotas** – The Rainforest Alliance has introduced a scheme called Smartlogging. This is a certification service that demonstrates that a logging company is working legally and in a sustainable way to protect the environment. The timber can be tracked from where it is felled to its final export destination and its use in timber products.

- **Replanting** – Areas of forest that have been cleared of trees are replanted, or equivalent numbers of trees are planted elsewhere when the land has been used for another purpose.

Conservation of fish stocks

- **Education** – This helps commercial fishing businesses to understand the effects of their methods, which may be harmful to fish stocks.

Also, they can learn that if they protect the environment, the fish can survive, grow well, breed and provide them with a living for the future.

- **Closed seasons** – These are parts of the year when it is illegal to fish for some species in an area of water. This allows the fish to spawn and mature, so they can breed successfully.
- **Protected areas** – These are clearly defined geographical spaces that are managed to protect the fish stocks in them. By preventing the capture and removal of fish, the stocks are allowed to rebuild.
- **Control of net types and mesh size** – This is to ensure that fish are not caught randomly, and that undersized fish can escape capture.
- **Legal quotas** – In Europe, the Common Fisheries Policy is used to set quotas for fishing, to manage fish stocks and to help protect species that are becoming endangered through over-fishing. Quotas are set for each species of fish taken commercially, and also for the size of fish. This is to allow fish to reach breeding age and to maintain or increase their populations.
- **Monitoring** – This involves recording commercial fish catches so that estimates of the fish populations can be made. Quotas are then adjusted to maintain the stocks.

Endangering species and causing their extinction

Many species of animal and plant are becoming endangered or are in danger of extinction. This is because of factors such as climate change, habitat destruction, hunting, over-harvesting, pollution and the introduction of other species.

Conservation of species

Conservation can sometimes be achieved by the monitoring and protection of species. Many organisations monitor species numbers so that conservation measures can be taken if they decline significantly.

The conservation of habitats is equally as important as the conservation of individual species. If habitats are lost, so are the species that live in them, so habitat destruction poses the greatest threat to the survival of species. A habitat may be conserved by:

- using laws to protect species and the habitat
- using wardens to protect species and the habitat
- reducing or controlling public access to the habitat
- controlling factors, such as water drainage and grazing, that may otherwise contribute to destruction of the habitat

Education plays an important role in helping local communities to understand why species need to be conserved.

Provided a species has not become totally extinct, it may be possible to boost its numbers by breeding in captivity and releasing the animals back into the environment.

Seed banks are a way of protecting plant species from extinction. They include seed from food crops and rare species, and act as gene banks.

Conservation programmes

Conservation programmes are set up for a number of reasons, which are outlined below.

Maintaining or increasing biodiversity

Extinction results in a reduction in biodiversity. This reduces the stock of hereditary material that could otherwise be used, for example, for crop improvement or as a source of new medicines. Gene banks have been set up to preserve a wide range of plants by storing their seeds. However, the best way of preserving the full range of genes is to keep plants growing in their natural environments.

Reducing extinction

Conservation programmes strive to prevent extinction. Once a species becomes extinct, its genes are lost forever, so we are also likely to deprive the world of genetic resources. We could deprive ourselves of the beauty and diversity of species, as well as potential sources of valuable products such as drugs.

Protecting vulnerable ecosystems

Conservation programmes are often set up to protect threatened habitats so that rare species living there are not endangered. There are a number of organisations involved with habitat conservation, such as the Worldwide Fund for Nature.

Maintaining ecosystem functions

There is a danger of destabilising food chains if a single species in that food chain is removed. For example, in lakes containing pike as the top predators, over-fishing can result in smaller species of carnivorous fish, such as minnows, increasing in numbers. They eat zooplankton. If the minnows eat the majority of the zooplankton population, it leaves no herbivores to control algal growth, which can lead to eutrophication. To prevent such an event happening, the ecosystem needs to be maintained by controlling the numbers of top predators removed or by regular restocking.

Ecosystems can also become unbalanced if the nutrients they rely on are affected in some way. This may be due to the unregulated removal of materials that affect food chains indirectly because of changes to nutrient cycles.

Humans are affecting ecosystems on a large scale because of the growth in the population and changing patterns of consumption. About 40% of the Earth's land surface area is taken over by some form of farmed land. Crops are grown for food, extraction of drugs (both legal and illegal) and the manufacture of fuel. Crop growth has major impacts on ecosystems, causing the extinction of many species and reducing the gene pool.

Use of AI and IVF in captive breeding programmes

Artificial insemination (AI) and in vitro fertilisation (IVF) are techniques used to improve fertility rates in captive breeding programmes.

AI involves collecting sperm samples from the male animal, and then artificially introducing them into a female's reproductive system to fertilise her eggs. The sperm can be used immediately, or frozen and stored. Being able to store the samples means that the sperm can be sent to other centres where captive breeding programmes are being run, increasing the genetic diversity of species.

IVF involves fertilisation being carried out in laboratory glassware. It is used in captive breeding programmes in species where females may not breed naturally but are still able to produce viable eggs, or the males may not produce adequate amounts of viable sperm. The female may be given fertility drugs, which cause her ovaries to release several mature ova at the same time. These ova are collected and then mixed with the male's seminal fluid and watched under the microscope to see if cell division takes place. One or more of the dividing zygotes are then introduced to the female's uterus.

Risks to a species of decreasing population size

The population of a species may decrease in size because of an increase in predation, disease, shortage of food or emigration (see Chapter 19). If the population is large, then there will be a minimal reduction in the genetic variation present. However, a smaller population would be affected because the genetic variation in the gene pool would be reduced. This would affect the ability of the species to cope with environmental change, and put it at greater risk of extinction.

Exam-style questions

1 a State how each of the following has enabled humans to increase food production:
   i chemical fertilisers [1]
   ii insecticides [1]
   iii herbicides [1]
   b State two other ways in which food production has been increased. [2]
2 a Describe the meaning of the term *monoculture*. [2]
   b Describe one advantage and one disadvantage of:
   i large-scale monoculture of crop plants [2]
   ii intensive livestock production [2]
3 a Define the term *biodiversity*. [2]
   b Deforestation is an example of habitat destruction.
   i State three reasons for habitat destruction. [3]
   ii One undesirable effect of deforestation is a reduction in biodiversity. Outline three other undesirable effects of deforestation. [3]
4 Figure 20.3 shows the area of tropical rainforest deforested annually in five different countries, labelled A to E.
   a i Which of the countries shown has the largest area deforested annually? [1]
   ii Which of the countries shown has 600 000 hectares of rainforest removed each year? [1]

   iii In another country (F), 550 000 hectares are deforested annually. Plot this on a copy of the figure. [1]
   b i Country E has a total of 9 000 000 hectares of tropical rainforest remaining. How long will it be before it is all destroyed, if the present rate of deforestation continues? [1]

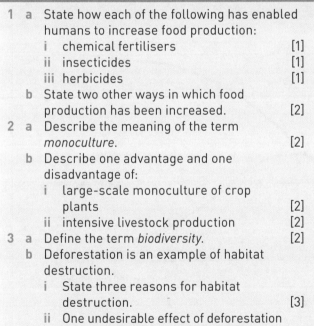
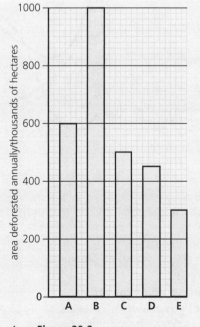

▲ **Figure 20.3**

ii State two reasons why tropical rainforests are being destroyed by humans. [2]

iii After deforestation has taken place, soil erosion often occurs rapidly. Suggest two ways in which this may occur. [2]

c Tropical rainforests reduce the amount of carbon dioxide and increase the amount of oxygen in the atmosphere. Explain why both these occurrences are important to living organisms. [2]

5 a Define the term *sustainable resource*. [2]

b Outline three ways in which endangered species can be conserved. [3]

6 Figure 20.4a shows part of a river into which sewage is pumped. The river water flows from W to Z, with the sewage being added at X. Some of the effects of adding sewage to the river are shown in Figure 20.4b.

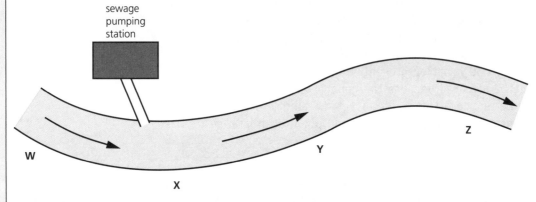

▲ Figure 20.4a

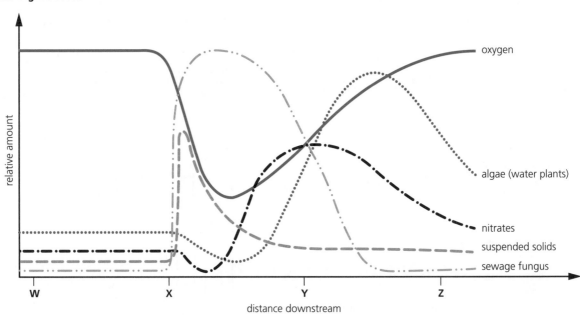

▲ Figure 20.4b

a Describe the changes from W to Z in the levels of:
i nitrates [2]
ii suspended solids [2]

b Suggest why the level of oxygen:
i drops at X [1]
ii increases again towards Z [1]

c Suggest two reasons why levels of algae drop:
i when sewage is added to the river [2]
ii towards Z [2]

d A farm at Z used herbicides on the field next to the river. Suggest why this could cause further problems in the river. [1]

7 State and explain four ways in which fish stocks can be conserved. [8]

Key objectives

The objectives for this chapter are to revise:
- why bacteria are useful in biotechnology and genetic modification
- the role of anaerobic respiration in yeast during the production of ethanol for biofuels and during breadmaking
- the use of pectinase in fruit juice production
- the use of biological washing powders that contain enzymes
- the definition of genetic modification and examples of its use

- the use of lactase to produce lactose-free milk
- how fermenters are used in the production of insulin, penicillin and mycoprotein
- genetic modification using bacterial production of a human protein
- the advantages and disadvantages of genetically modifying crops

Key term

Term	Definition
Genetic modification	Changing the genetic material of an organism by removing, changing or inserting individual genes

Biotechnology and genetic modification

Biotechnology is the application of biological organisms, systems or processes to manufacturing and service industries. **Genetic modification** involves the transfer of genes from one organism to (usually) an unrelated species. Both processes often make use of bacteria because of their ability to make complex molecules (e.g. proteins) and their rapid reproduction rate.

Use of bacteria in biotechnology and genetic modification

Bacteria are useful in biotechnology and genetic modification because:

- they can be grown and manipulated without raising ethical concerns
- they have a genetic code that has the same basis as all other organisms, so genes from other animals or plants can be successfully transferred into bacterial DNA

Bacterial DNA is in the form of a circular strand, and also small circular pieces called **plasmids.** Scientists have developed techniques to cut open these plasmids and insert sections of DNA from other organisms into them. When the bacterium divides, the DNA in the modified plasmid is copied, including the 'foreign' DNA. This may contain a gene to make a particular protein.

Biotechnology

Biofuels

In Chapter 12, the anaerobic respiration of glucose to alcohol is described as a form of **fermentation**. Microorganisms that bring about fermentation are using the chemical reaction to produce energy, which they need for their living processes.

Yeast is encouraged to grow and multiply by providing nutrients such as sugar. An optimum pH and temperature are maintained for the yeast being cultured, and oxygen or air is excluded to maintain an anaerobic process. Ethanol is a waste product. Some countries produce ethanol in this way as a renewable source of energy (**biofuel**) for cars, replacing non-renewable petrol.

Bread

Yeast is used in breadmaking and brewing because of the products produced when it respires anaerobically (see Chapter 12). The yeast is mixed with water and sugar to activate it. This mixture is then added to flour to make dough. This is left in a warm place to rise. The dough rises because the yeast is releasing carbon dioxide, which gets trapped in the dough. A warm temperature is important because respiration is controlled by enzymes (see Chapter 5). The dough is then cooked. The high temperature kills the yeast, and any ethanol formed evaporates. Air spaces are left where the carbon dioxide was trapped. This gives the bread a light texture.

Fruit juice production

When pectinase is added to fruit tissue, it breaks down the tissue, releasing sugars in solution and making the liquid extract transparent (clarifying it). The process occurs faster in warm conditions than in cold conditions because it is controlled by an enzyme.

Biological washing powders

Biological washing powders can contain protease and lipase to remove protein stains and fat/grease from clothes more effectively than detergent alone. The enzymes break down the proteins and fats in the clothes to amino acids, fatty acids and glycerol. These are smaller, soluble molecules, which can escape from the clothes and dissolve in the water.

The relatively low temperatures at which enzymes work best make biological washing powder more efficient because this saves energy (no need to boil water). However, if the temperature is too high the enzymes will be denatured.

Revision activity

Make a mind-map for biotechnology. For each of the applications (breadmaking etc.) include the main features of the application. Figure 21.1 shows how you could start to plan your diagram.

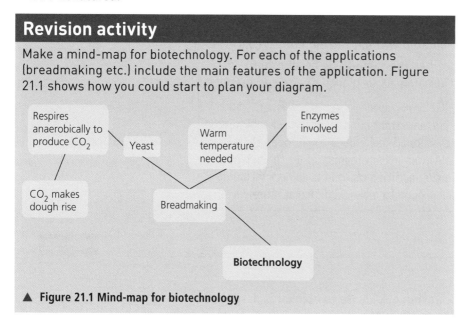

▲ **Figure 21.1 Mind-map for biotechnology**

Lactose-free milk

Lactose is a type of sugar found in milk and dairy products. Some people suffer from **lactose intolerance**, a digestive problem in which the body does not produce enough of the enzyme lactase. As a result, the lactose remains in the gut, where it is fermented by bacteria, causing symptoms such as flatulence (wind), diarrhoea and stomach pains. Many foods contain dairy products, so people with lactose intolerance cannot eat them, or suffer the symptoms described above. However, lactose-free milk is now produced using the enzyme lactase.

Lactase can be produced on a large scale by fermenting yeasts or fungi. The fermentation process is shown in Figure 21.2.

A simple way to make lactose-free milk is to add lactase to milk. The enzyme breaks down lactose sugar into two monosaccharide sugars: glucose and galactose. Both can be absorbed by the intestine.

An alternative large-scale method is to immobilise lactase on the surface of beads. The milk is then passed over the beads and the lactose sugar is broken down. This method avoids having the enzyme molecules in the milk because they remain on the beads.

How fermenters are used in the production of penicillin

A fermenter (Figure 21.2) is a large, sterile container with a stirrer, a pipe to add feedstock (molasses or corn-steep liquor) and air pipes to blow air into the mixture.

The fungus *Penicillium* is added and the liquid is maintained at around 26°C and a pH of 5–6. Sterile conditions are essential to prevent 'foreign' bacteria or fungi getting into the system, as they can completely disrupt the process. As the nutrient supply reduces, the fungus begins to secrete antibiotics into the medium. The nutrient fluid containing the antibiotic is filtered off and the antibiotic is extracted by crystallisation or other methods.

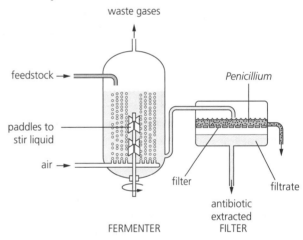

▲ **Figure 21.2 Principles of antibiotic production using a fermenter**

Commercial production of insulin

Insulin can also be produced in large quantities using fermenters. Bacteria are genetically modified to carry the human insulin gene and grown in a fermenter. They respire aerobically, so air is pumped into the fermenter. Other conditions, such as nutrient levels, temperature, pH and moisture, are maintained at optimum levels so that the bacteria grow and reproduce rapidly. The nutrients are then reduced to stimulate the bacteria to produce the insulin.

Mycoprotein

Mycoprotein is a protein-rich meat-substitute extracted from fungi. Its manufacture has been developed so that it can be made commercially. It is fermented in a similar way to antibiotics and enzymes, using glucose and salts as the feedstock. One commercially available mycoprotein product is called Quorn®.

Table 21.1 outlines the conditions that need to be maintained in fermenters during these processes.

▼ **Table 21.1 Conditions that need to be maintained in a fermenter**

Condition	Reason
Temperature	The process involves enzymes. The enzymes in *Penicillium* work best at 26°C. Heat is generated during fermentation, so the mixture needs to be cooled.
pH	The fungus *Penicillium* requires a slightly acidic environment of pH 5–6 to ferment at its optimum rate.
Oxygen	Sterilised air is blown into the mixture through air pipes and the mixture is stirred to aerate it. This allows healthy growth of the antibiotic or enzyme.
Nutrient supply	This depends on what is being manufactured, but for penicillin the feedstock is molasses or corn-steep liquor. This allows the *Penicillium* to grow and reproduce at its optimum rate. The nutrient supply is reduced when the population is large, to stimulate it to secrete antibiotics.
Waste products	These depend on what is being manufactured, but for penicillin they are waste nutrients with bacterial residue. Gases given off may include carbon dioxide. These need to be removed as they are produced because the rate of growth of *Penicillium* becomes limited by the presence of metabolic waste products.

Genetic modification

REVISED

You need to learn the definition of **genetic modification**, given at the start of this chapter, and be able to give examples. Some are listed below:

- The production of human insulin – the human insulin gene is inserted into bacteria. Human insulin does not trigger allergic reactions in the way that animal insulin can, and is acceptable to people with a range of religious beliefs.

- The insertion of genes into crop plants to give them resistance to herbicides (weedkillers) – this enables the farmer to spray the crop to kill weeds, without damaging the crop, and may reduce the use of herbicides.

- The insertion of genes into crop plants to give them resistance to insect pests – the gene enables the plant to produce a poison that makes it resistant to attack by insect larvae.

- The insertion of genes into crop plants to provide additional vitamins – golden rice is a variety of rice that has had a gene for beta-carotene (a precursor of vitamin A) inserted. Golden rice is grown particularly in countries where vitamin A deficiency is a problem and where rice is a staple food. This deficiency often leads to blindness.

The advantages and disadvantages of genetically modifying crops are outlined in Table 21.2

▼ **Table 21.2 Advantages and disadvantages of genetically modifying crops**

Advantages	Disadvantages
The aim of most genetic modification is to increase yields through the insertion of genes giving crops herbicide resistance and insect pest resistance; genetically modified (GM) maize has resistance to pests and herbicides, while GM soya has been modified to make it herbicide resistant	The vectors for delivering recombinant DNA contain genes for antibiotic resistance; if these managed to get into potentially harmful bacteria, it might make them resistant to antibiotic drugs
It is possible to improve the protein, mineral or vitamin content of food – for example, golden rice has a gene enabling it to produce a precursor of vitamin A, while GM soya has an increased nutritional value	GM food could contain pesticide residues or substances that cause allergies
	The precursor of vitamin A in golden rice could change into other, toxic chemicals once eaten
It is possible to improve the keeping qualities of some products – for example, the storage properties of GM soya have been improved through modification of its fat molecules, using inserted genes, while GM tomatoes have had a gene deleted that is responsible for fruit softening, extending their storage life	There is the risk of a reduction in biodiversity as a result of the introduction of GM species
	Subsistence farmers could be tied to large agricultural suppliers that may then manipulate seed prices

Using genetic modification to put human insulin genes into bacteria

Figure 21.3 shows this process. The steps, numbered on the diagram, are as follows:

1 Human cells with genes for healthy insulin are selected.
2 A chromosome (which is a length of DNA) is removed from the cell and isolated.
3 The section of DNA representing the insulin gene is cut from the chromosome using **restriction enzymes**. The DNA has sticky ends.
4 A suitable bacterial cell is selected. Some of its DNA is in the form of circular **plasmids**.
5 All the plasmids are removed from the bacterial cell.
6 The plasmids are cut open using the same restriction enzymes, forming complementary sticky ends.
7 The human insulin gene is inserted into the plasmids using **DNA ligase** enzyme, forming **recombinant plasmids**.
8 The plasmids are returned to the bacterial cells (only one is shown in the diagram).
9 The bacterial cells are allowed to reproduce in a fermenter. All the cells produced contain plasmids that express the human insulin gene. The bacteria can now be used to produce human insulin on a commercial scale.

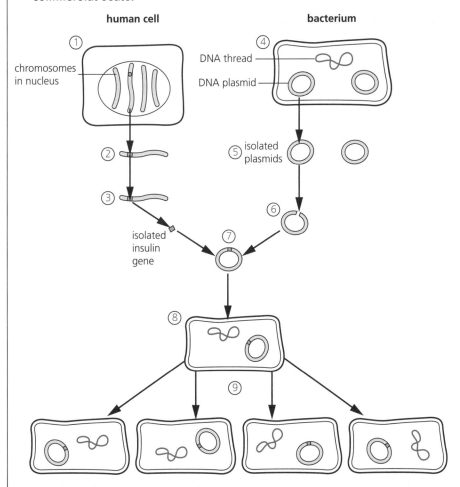

▲ **Figure 21.3 Using genetic modification to put human insulin genes into bacteria**

Sample question

Scientists are planning the use of a genetically modified virus to destroy a population of an amphibian called the cane toad, which is getting out of control in Australia.

a Define the term *genetic modification*. [2]

b State the part of the virus that would carry the modified genetic material. [1]

Student's answer

a *Genetic modification is changing the genetic material* ✔ *in an organism by inserting genes* ✔ *from a different species.*

b *The nucleus* ✘

Teacher's comments

The definition given in part a is excellent. In part b, the student does not know that a virus lacks a nucleus – it contains DNA or RNA in the form of a strand.

Exam-style questions

1 State why bacteria are very useful organisms in the process of genetic modification. [2]

2 a Describe the meaning of the term *biotechnology*. [2]

   b State two factors in common between breadmaking and the production of ethanol for biofuels. [2]

   c Explain why:
      i   bread rises [2]
      ii bread does *not* contain ethanol [2]

3 a State two roles of pectinase in fruit juice production. [2]

   b Silk is a protein. Suggest why clothes made of silk should *not* be washed using biological washing powder. [2]

4 a Describe the role of lactase in the production of lactose-free milk. [2]

   b State two advantages of immobilising lactase on the surface of beads in the large-scale production of lactose-free milk. [2]

5 a Define the term *genetic modification*. [2]

   b i   Outline two examples of genetic modification. [4]
      ii For each example you have outlined, state an advantage. [2]

6 a State the roles of enzymes in the process of inserting human insulin genes into bacteria. [3]

   b Suggest why it is important that the cut plasmids have complementary sticky ends. [1]

Index

Note: page numbers in **bold** refer to the location where a key definition is *first* defined.